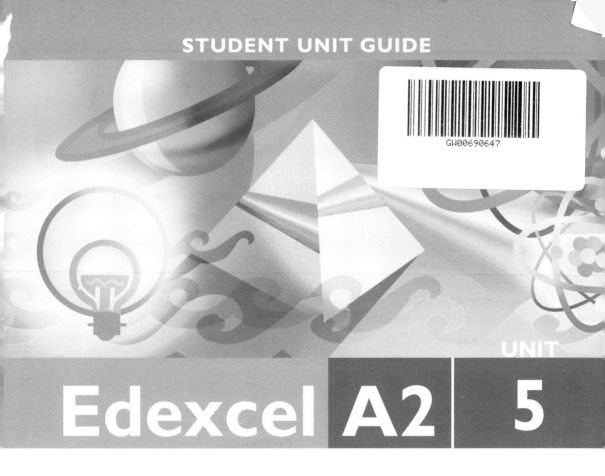

Edexcel **A2**

UNIT

5

Physics

Physics from Creation to Collapse

Mike Benn

Philip Allan Updates, an imprint of Hodder Education, an Hachette UK company, Market Place, Deddington, Oxfordshire OX15 0SE

Orders

Bookpoint Ltd, 130 Milton Park, Abingdon, Oxfordshire OX14 4SB
tel: 01235 827827
fax: 01235 400401
e-mail: education@bookpoint.co.uk
Lines are open 9.00 a.m.–5.00 p.m., Monday to Saturday, with a 24-hour message answering service. You can also order through the Philip Allan Updates website: www.philipallan.co.uk

© Philip Allan Updates 2010

ISBN 978-0-340-94828-6

First printed 2010
Impression number 5 4 3 2
Year 2014 2013 2012 2011

This guide has been written specifically to support students preparing for the Edexcel A2 Physics Unit 5 examination. The content has been neither approved nor endorsed by Edexcel and remains the sole responsibility of the author.

Typeset by Pantek Arts Ltd, Maidstone, Kent
Printed in India

Hachette UK's policy is to use papers that are natural, renewable and recyclable products and made from wood grown in sustainable forests. The logging and manufacturing processes are expected to conform to the environmental regulations of the country of origin.

Contents

Introduction

■ ■ ■

Content Guidance

■ ■ ■

Questions and Answers

Introduction

About this guide

This guide is one of a series covering the Edexcel specification for AS and A2 physics. It offers advice for the effective study of **Unit 5: Physics from Creation to Collapse**. Its aim is to help you to *understand* the physics — it is not intended as a shopping list that enables you to cram for the examination.

The guide has three sections:

- **Introduction** — this gives brief guidance on approaches and techniques to ensure that you answer the examination questions in the best way that you can.
- **Content Guidance** — this is not intended to be a detailed textbook. It offers guidance on the main areas of the content of Unit 5, with an emphasis on worked examples. These examples illustrate the types of question that you are likely to come across in the examination.
- **Questions and Answers** — this contains two unit tests, presented in a format close to that of an actual Edexcel examination and using questions similar to those in recent past papers, to give the widest possible coverage of the unit content. Answers are provided — in some cases, distinction is made between responses that might have been written by an A-grade candidate and those typical of a C-grade candidate. Common errors made by candidates are also highlighted so that you, hopefully, do not make the same mistakes!

Understanding physics requires time and effort. No one suggests that physics is an easy subject, but even students who find it difficult can overcome their problems by the proper investment of time.

A deep understanding of physics can develop only with experience, which means time spent thinking about physics, working with it and solving problems. This book provides you with a platform for doing this. If you try all the worked examples and the unit tests *before* looking at the answers, you will begin to think for yourself and develop the necessary techniques for answering examination questions effectively. In addition, you need to *learn* all the basic formulae, definitions and experiments. Thus prepared, you will be able to approach the examination with confidence.

The specification

The specification outlines the physics that will be examined in the unit tests and describes the format of those tests. This is not necessarily the same as your teachers might choose to teach (or what you may choose to learn).

The purpose of the book is to help you with the Unit 5 test, but don't forget that what you are doing is learning *physics*. The specification can be obtained from Edexcel, either as a printed document or from the web at **www.edexcel.com**.

The unit test

Unit Test 5 is a written paper of duration 1 hour 35 minutes. It carries a total of 80 marks. There are ten objective (multiple-choice) questions, each worth a single mark, and further short and long questions worth between 3 and 13 or more marks. Unit 5 counts for 40% of the A2 marks or 20% of the total A-level marks.

Questions will assume that Units 1 and 2 have been studied, but will not examine the content again in detail — for example, knowledge of oscillations and waves will be expected in the study of simple harmonic motion.

Unit Test 5 will examine the objectives AO1 (knowledge and understanding), AO2 (application of knowledge and understanding, synthesis and evaluation) and AO3 (how science works). There is a slight change of emphasis towards AO2 compared with AS papers. The questions are less structured than at AS. Quantities may not be given in base units so, for example, you may have to convert measurements in nano-metres and millimetres into metres before substituting into the appropriate equation.

Command terms

Examiners use certain words that require you to respond in a particular way. You must be able to distinguish between these terms and understand exactly what each requires you to do. Some frequently used commands are described below.

- **State** — the answer should be a brief sentence giving the essential facts; no explanation is required (nor should you give one).
- **Define** — you can use a *word equation*; if you use *symbols* you must state what each symbol represents.
- **List** — simply give a series of words or terms; there is no need to write sentences.
- **Outline** — a logical series of bullet points or phrases will suffice.
- **Describe** — for an experiment a diagram is essential; then state the main points concisely (bullet points can be used).
- **Draw** — diagrams should be drawn in section, neatly and *fully labelled* with all measurements clearly shown, but don't waste time — remember that this is not an art exam.
- **Sketch** — usually a graph is called for, but graph paper is not necessary (although a grid is sometimes provided); axes must be labelled and include a scale if numerical data are given; the origin should be shown if appropriate, and the general shape of the expected line or curve should be drawn clearly.
- **Explain** — use correct physics terminology and principles; the amount of detail in your answer should reflect the number of marks available.

- **Show that** — usually a value is provided (to enable you to proceed with the next part of the question) and you have to demonstrate how this value can be obtained; you should show all your working and state your result to more significant figures than the given value contains (to prove that you have actually done the calculations).
- **Calculate** — show all your working and include *units* at every stage; the number of significant figures in your answer should reflect the given data, but you should keep each stage with more significant figures in your calculator to prevent excessive rounding.
- **Determine** — you will probably have to extract some data, often from a graph, in order to perform a calculation.
- **Estimate** — this means doing a calculation in which you have to make a sensible assumption, possibly about the value of one of the quantities. Think: does your assumption lead to a reasonable answer?
- **Suggest** — there is often no single correct answer; credit is given for sensible reasoning based on correct physics.
- **Discuss** — you need to sustain an argument, giving evidence for and against based on your knowledge of physics and possibly using appropriate data to justify your answer.

You should pay particular attention to diagrams, graph sketching and calculations. Candidates often lose marks by failing to label diagrams properly, by not giving essential numerical data on sketch graphs, and by not showing all the working or by omitting units in calculations.

Revision

The purpose of this introduction is not to provide you with an in-depth guide to revision techniques — you can find many books on study skills if you feel you need more help in preparing for examinations. There are, however, some points worth mentioning that will help you when you are revising the physics A-level material:

- Familiarise yourself with what you need to know — ask your teacher and look through the specification.
- Make sure you have a good set of notes — you can't revise properly from a textbook.
- Learn all the equations indicated in the specification and become familiar with the formulae that will be provided in the examination (at the end of each question paper) so that you can find them quickly and use them correctly.
- Make sure that you learn definitions thoroughly and in detail — for example, simple harmonic motion requires that the acceleration of an oscillating particle is *directly proportional to its displacement from a point* and is always *directed towards that point*.
- Be able to describe (with diagrams) the basic experiments referred to in the specification.
- Make revision active by writing out equations and definitions, drawing diagrams, describing experiments and by performing lots of calculations.

Content
Guidance

This section is a guide to the content of **Unit 5: Physics from Creation to Collapse**. The unit is split into four main topics: **Thermal energy**, **Nuclear decay**, **Oscillations** and **Astrophysics and cosmology**.

Thermal energy
- Specific heat capacity of solids and liquids.
- Internal energy in terms of the kinetic and potential energies of molecules — the concept of absolute zero of temperature in terms of the average kinetic energy of molecules.
- Ideal gases — the gas laws including the use of the equation of state for an ideal gas ($pV = NkT$); kinetic theory — the use and understanding of $\frac{1}{2}m\langle c^2 \rangle = \frac{3}{2}kT$.

Nuclear decay
- Nuclear radiation; properties of alpha, beta and gamma radiations (including ionising ability and penetrating power); background radiation.
- Expressions for exponential decay ($\frac{dN}{dt} = -\lambda N$ and $N = N_0 e^{-\lambda t}$); half-life.
- Applications of radioactive materials; ethical and environmental issues.

Oscillations
- Simple harmonic motion, including the relationships for acceleration in terms of displacement ($a = -kx$) and time ($a = -A\omega^2 \cos t$); the expressions for velocity and displacement in terms of time for a simple harmonic oscillator ($v = A\omega \sin t$ and $x = A\cos\omega t$).
- Energy in simple harmonic oscillators; free, forced and damped oscillations; resonance.

Astrophysics and cosmology
- Newton's law of gravity ($F = \frac{Gm_1 m_2}{r^2}$); gravitation field strength ($g = \frac{Gm}{r^2}$); similarities and differences between electric and gravitational fields.
- Relationship between luminosity and flux for stars ($F = \frac{L}{4\pi d^2}$); Stefan–Boltzmann law ($L = 4\pi r^2 \sigma T^4$); Wien's law ($\lambda_{max} T = 2.898 \times 10^{-3}\, \text{m K}$)
- Hertzsprung–Russell diagram; the life cycle of stars; evidence for the expanding universe ($z = \frac{\Delta\lambda}{\lambda} \approx \frac{\Delta f}{f} \approx \frac{v}{c}$); Hubble's law ($v = H_0 d$); dark matter.
- The concept of nuclear binding energy; use of the expression $\Delta E = c^2 \Delta m$; the processes of nuclear fission and nuclear fusion; mechanism of nuclear fusion in stars, and the conditions necessary for fusion to occur.

Quantity algebra

Quantity algebra is used throughout. This involves putting into an equation the units for each quantity. For example, to calculate the pressure of 2 moles of an ideal gas occupying a volume of $50\,cm^3$ at a temperature of 27°C, we would write:

$$pV = nRT \Rightarrow p = \frac{nRT}{V} = \frac{2 \times 8.3\,J K^{-1} mol^{-1} \times (273+27)\,K}{50 \times 10^{-6}\,m^3} = 1.0 \times 10^8\,Pa$$

Although Edexcel does not require you to use quantity algebra, you are strongly advised to do so because of the following advantages:

- It acts as a reminder to substitute consistent units. In this example, $50\,cm^3$ is written as $50 \times 10^{-6}\,m^3$, and °C is converted to K.
- It allows you to check that the units of the answer are correct. In this example, Jm^{-3} is equivalent to $(Nm)m^{-3}$, i.e. Nm^{-2} or Pa.

Thermal energy

Thermal energy, or heat, is the transfer of energy from a hot (high temperature) body to a cold body.

In this section we will examine the effect of heat on different materials, and look into the behaviour of gases by studying the equation of state of an ideal gas and the kinetic theory of gases.

- **Heat** is energy transfer from a body or region of high temperature to one of lower temperature.
- **Internal energy** is the sum of the kinetic energy of the molecules within a body and the potential energy of the bonds between the molecules.

It is a common error to refer to the internal energy of a body as the 'heat contained within it'.

Specific heat capacity

When a hot object is in thermal contact with a cooler body, there will be a net flow of energy from the region of high temperature to the lower one. If no energy is added to or removed from the system, the temperature of the hot body will fall and that of the cold one will increase. The rise or fall in temperature is dependent on the energy transferred, the masses of the bodies and the nature of the molecular structure of the material.

- **Specific heat capacity**, c, is the energy transfer needed to change the temperature of 1 kilogram of a substance by 1 kelvin:

$$c = \frac{\Delta E}{m\Delta\theta} \qquad \text{units: } J\,kg^{-1}\,K^{-1}$$

$$\Delta E = mc\Delta\theta$$

Later in this section, the relationship between temperature and the average kinetic energy of molecules will be developed. A kilogram of water contains more molecules than there are atoms in 1 kilogram of copper. The same increase in internal energy will therefore be shared by fewer particles in the copper, and so the average energy of each will be larger than for water. It follows that more energy is needed to raise the temperature of 1 kilogram of water by 1 kelvin than is needed for 1 kilogram of copper.

Worked example

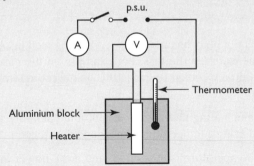

A student set up this apparatus to measure the specific heat capacity of aluminium.

A heater was placed in the block and connected to a power supply. The initial temperature of the block was measured and the supply was switched on. After 10 minutes the power was switched off and the final temperature was recorded. The student's results are given below:

mass of block, $m = 0.800\,\text{kg}$

voltage, $V = 12.0\,\text{V}$

current, $I = 3.2\,\text{A}$

Initial temperature of block = 18.0°C; final temperature = 46.0°C

(a) Show that the student's results give a value of about $1000\,\text{J}\,\text{kg}^{-1}\,\text{K}^{-1}$ for the specific heat capacity of aluminium.

(b) The accepted value of the specific heat capacity of aluminium is $960\,\text{J}\,\text{kg}^{-1}\,\text{K}^{-1}$. Suggest the likely explanation of why the student's value is higher than this.

Answer

(a) $\Delta E = VIt = mc\Delta\theta$

$$\Rightarrow c = \frac{VIt}{m\Delta\theta}$$

$$= \frac{12.0\,\text{V} \times 3.2\,\text{A} \times 600\,\text{s}}{0.800\,\text{kg} \times 28°\text{C}}$$

$$= 1030\,\text{J}\,\text{kg}^{-1}\,\text{K}^{-1}$$

Note: You are expected to know the expression for electrical energy used in Unit 2.

(b) As the block gets hotter, heat will be lost to the surroundings, so not all of the electrical energy from the heater is transferred to the aluminium. The value used for ΔE in the calculation is therefore higher than it should be, and so the calculated value of c will also be too big.

In experiments to determine specific heat capacities, heat energy, ΔQ, may be lost to or gained from the surroundings.

$$\Delta E = mc\Delta\theta \pm \Delta Q$$

To minimise this effect, a body being heated needs to be well insulated, or an allowance made to compensate for the loss. For example, in the case of a body being heated, it could be initially cooled to, say, 10°C below room temperature and then heated to 10°C above.

Another method of eliminating heat loss is to plot a temperature against time graph (this can be achieved using a temperature sensor and data logger). The gradient of the graph $\left(\dfrac{\Delta\theta}{\Delta t}\right)$ at room temperature is measured, and if the power input is known, the specific heat capacity is found using:

$$P = \frac{\Delta E}{\Delta t}$$

$$= mc\frac{\Delta\theta}{\Delta t} + \frac{\Delta Q}{\Delta t}$$

If the energy is supplied from an electrical source then $P = VI$, and at room temperature no heat is lost or gained from the body, so $\dfrac{\Delta Q}{\Delta t}$ is zero.

Steady state conditions

Energy can be transferred continuously to moving fluids to achieve a fixed rise in temperature. For example, a hot-water boiler may supply heat to cold water so that it leaves the boiler at a constant temperature. When this occurs, **steady state** conditions have been reached and the temperature rise will depend on the power input, the specific heat capacity of water and the rate of flow $\left(\dfrac{\Delta m}{\Delta t}\right)$:

$$P = \frac{\Delta m}{\Delta t}c\Delta\theta + \frac{\Delta Q}{\Delta t}$$

The gas laws

When gases are heated, cooled, compressed or expanded, there are interrelated variations in pressure, volume and temperature. Observations of these variations led to three 'laws' that applied to most gases over a range of temperatures and pressures.

- **Boyle's law** states that for a fixed mass of gas at constant temperature, the pressure is inversely proportional to the volume: $p \propto \dfrac{1}{V}$.
- **Charles' law** states that for a fixed mass of gas at constant pressure, the volume is proportional to the temperature in kelvins: $V \propto T$.
- **The pressure law** states that for a fixed mass of gas at constant volume, the pressure is directly proportional to the temperature in kelvins: $p \propto T$.

You will be expected to know details of experiments to test these relationships. Data logging devices using pressure, temperature and displacement sensors are recommended.

Graphical representations of the gas laws are shown below.

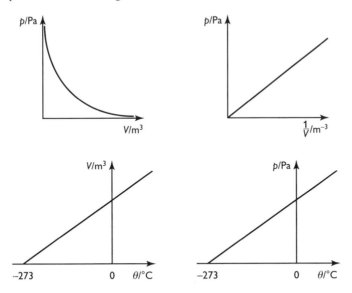

These laws hold for most gases at low pressure and high temperature, but at high pressures and low temperatures they do not apply, and some gases may even liquefy.

Ideal gases

An ideal gas obeys Boyle's law at *all* temperatures, meaning that it will *never* liquefy, and that there are *no* intermolecular forces. The internal energy is therefore solely kinetic.

→ maintain constant pressure.

An **ideal gas** is defined by the 'equation of state':

$pV = nRT$ **for n moles of a gas**

R is the universal molar gas constant — its value is $8.31\,\text{J}\,\text{K}^{-1}\,\text{mol}^{-1}$. The Edexcel specification defines an ideal gas in terms of the number of molecules, N, of the gas. One mole of a gas contains Avogadro's number (N_A) of particles, so the number of moles for N molecules will be $\dfrac{N}{N_A}$. The **equation of state** can be written:

$$pV = \frac{N}{N_A} RT$$

$$= NkT$$

The constant, $k\left(=\dfrac{R}{N_A}\right)$, is the **Boltzmann constant**, which has a value of $1.38 \times 10^{-23}\,\text{J}\,\text{K}^{-1}$. The equation of state for an ideal gas is given in terms of the number of molecules and the Boltzmann constant in the data sheet at the end of the examination paper, together with the value of k.

Worked example

An ideal gas is contained in a volume of $1.2 \times 10^{-3} \, \text{m}^{-3}$. Calculate the number of molecules within the container if the pressure of the gas is 2.0×10^5 Pa and the temperature is $57°C$.

Answer

$$pV = NkT \Rightarrow N = \frac{pV}{kT}$$

$$N = \frac{2.0 \times 10^5 \, \text{Pa} \times 1.2 \times 10^{-3} \, \text{m}^3}{1.38 \times 10^{-23} \, \text{JK}^{-1} \times (273 + 57) \, \text{K}}$$

$$= 5.3 \times 10^{22}$$

Note: Examiners usually give temperatures in °C; they must be converted to K for all ideal gas calculations.

Kinetic theory for an ideal gas

The kinetic model of a gas is described in terms of vast numbers of particles moving randomly in a container. The temperature of the gas is related to the speed of the particles, and the pressure on the walls of the container is produced by the momentum changes of the particles as they collide with the walls.

The following assumptions are made when applying the **kinetic theory**:

- There are no inter-particle forces except during collisions.
- The time spent during collisions is negligible compared to the time in free motion.
- Collisions are, on average, perfectly elastic.
- The collisions obey Newton's laws of motion.
- The volume occupied by the particles is negligible compared with the volume of the gas.

Using these assumptions it is possible to derive the expression:

$$pV = \frac{1}{3} Nm\langle c^2 \rangle$$

where m is the mass of each molecule and $\langle c^2 \rangle$ is the mean square value of the molecular speeds.

Combining this with the equation of state of an ideal gas given above gives:

$$\frac{1}{3} Nm\langle c^2 \rangle = NkT$$

$$\frac{1}{2} m\langle c^2 \rangle = \frac{3}{2} kT$$

This is an important expression, because it relates the average kinetic energy of the molecules to the temperature on the Kelvin scale. We can also see that there must be an **absolute zero** of temperature (0 K) when the kinetic energy of the molecules is zero.

> **Worked example**
> Calculate the root mean square speed of nitrogen molecules in air at a
> temperature of 20°C (mass of nitrogen molecule = 4.6×10^{-26} kg).
>
> *Answer*
>
> $$\frac{1}{2}m\langle c^2 \rangle = \frac{3}{2}kT$$
>
> $$\Rightarrow \langle c^2 \rangle = \frac{3kT}{m}$$
>
> $$\langle c^2 \rangle = \frac{3 \times 1.38 \times 10^{-23}\,\mathrm{J\,K^{-1}} \times (273 + 20)\,\mathrm{K}}{4.6 \times 10^{-26}\,\mathrm{kg}}$$
>
> $$= 2.64 \times 10^5\,\mathrm{m^2\,s^{-2}}$$
>
> $$\sqrt{\langle c^2 \rangle} = 510\,\mathrm{m\,s^{-1}}$$

Nuclear decay

Most of the elements around us are stable. There are others with **unstable nuclei** that
decay with the emission of radiation and eventually reach a stable condition. You might
expect that, eventually, all nuclei would become stable, and no radioactive material
would remain. We are, however, exposed to a continuous **background radiation**.

Background radiation is the result of three main causes:

- **Cosmic rays** — produced by nuclear events in space. Most are filtered out by the
  atmosphere, but airline pilots receive significantly higher doses than most people.
- **Sources in the Earth** — rocks containing uranium or those which release radon
  gas into the atmosphere. Building materials, and even the food we eat, contain
  traces of radioactive isotopes.
- **Man-made sources** — luminous dials; by-products from nuclear reactors, nuclear
  weapons tests and nuclear medicine.

The level of background radiation varies from place to place. Regions with high counts
will exist around certain rock formations (usually granite), at high altitudes and where
weapons testing has taken place. It should be noted that most of the background radiation is due to natural events, with only about 3% due to human activities.

When measurements of radiation are made in the laboratory, it is important that the
background level is also measured and subtracted from the readings.

Alpha, beta and gamma radiation

Radiation from decaying nuclei takes the form of alpha (α) particles and beta (β) particles, often accompanied by high-energy photons which make up gamma (γ) rays.

- **Alpha particles** are helium nuclei, ^4_2He, often represented as $^4_2\alpha$.
- **Beta particles** are electrons or positrons emitted from the nucleus. They are represented by the symbols β^- (beta-minus) and β^+ (beta-plus); in nuclear equations the symbols $^0_{-1}\beta$, $^0_{+1}\beta$ or $^0_{-1}\text{e}$, $^0_{+1}\text{e}$ should be used.
- **Gamma rays** are photons of energy $\Delta E = hf$. They are released from an excited nucleus when it returns to its ground state.

When alpha and beta particles are emitted from a nucleus, the proton number of the residual nucleus is changed so that a new element is formed. Some examples of nuclear decays are given below:

- $^{238}_{92}\text{U} \rightarrow {}^{234}_{90}\text{Th} + {}^4_2\text{He} + \gamma$ (alpha and gamma)
- $^{234}_{90}\text{Th} \rightarrow {}^{234}_{91}\text{Pa} + {}^0_{-1}\beta + \gamma$ (beta-minus and gamma)
- $^{15}_8\text{O} \rightarrow {}^{15}_7\text{N} + {}^0_{+1}\beta$ (beta-plus)
- $^{99\text{m}}_{43}\text{Tc} \rightarrow {}^{99}_{43}\text{Th} + \gamma$ (gamma only)

Note: You are not expected to remember specific examples of decay equations in the examination. However, you may be presented with an unfamiliar decay equation with incomplete proton and nucleon numbers. The following simple rules can be applied:

- Alpha emission — proton number reduced by two, nucleon number reduced by four.
- Beta-minus emission — proton number increased by one, nucleon number unchanged.
- Beta-plus emission — proton number decreased by one, nucleon number unchanged.
- Gamma ray emission — no change in proton number or nucleon number.

The neutron number, N, against proton number, Z, graph for stable nuclei is shown below:

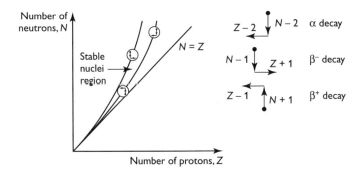

It can be seen that for higher proton numbers, the ratio of neutrons to protons increases for the stable nuclei. Unstable isotopes will lie outside this stability curve, and will decay in such a way that the daughter nucleus moves towards the stability region.

Worked example

For the following decays, complete the equations by adding the appropriate values of proton and nucleon numbers:

(a) $^{241}_{95}\text{Am} \rightarrow \text{Np} + \alpha + \gamma$

(b) $^{60}\text{Co} \rightarrow {}_{28}\text{Ni} + \beta^- + \gamma$

Answer

(a) $^{241}_{95}\text{Am} \rightarrow {}^{237}_{93}\text{Np} + {}^{4}_{2}\alpha + \gamma$

(b) $^{60}_{27}\text{Co} \rightarrow {}^{60}_{28}\text{Ni} + {}^{0}_{-1}\beta \rightarrow + \gamma$

Note: Sometimes the gamma ray is represented as ${}^{0}_{0}\gamma$ or *hf*, but it is usually given without proton and nucleon numbers.

Properties of alpha, beta and gamma radiation

Alpha particles are much bigger than beta particles ($m_\alpha \approx 2000 m_\beta$) and carry twice the charge. This means that they will readily displace electrons from atomic orbits and produce heavy ionisation. **Beta particles** have the same mass as atomic electrons and, despite having relatively high energy, they will displace fewer electrons and so will be less ionising than alpha particles.

The highly ionising alpha particles will lose their energy over a much shorter distance than beta particles, and so their penetrating power is very low — alpha particles have a range of only about 4 cm in air, and are absorbed by a thin sheet of paper. Beta particles can pass through several metres of air and through paper, but are stopped by a few millimetres of aluminium.

Gamma radiation is absorbed by a different mechanism to alpha and beta particles. Gamma rays do not produce a trail of continuous ionisation but transfer all, or most, of their energy to electrons in a single interaction. The degree of ionisation is much less than either alpha or beta particles and so several centimetres of lead are needed to absorb the rays.

The nature of the radiation from a given source can be deduced from a simple absorption test. First, place a piece of paper between the source and a suitable detector (e.g. Geiger tube). If no radiation is detected then the source is purely an alpha emitter; if there is no reduction in the reading then the radiation has no alpha particles but may consist of both beta and gamma. To find out if the radiation is beta or gamma, or a combination, a sheet of aluminium about 2 mm thick is placed between the source and detector. Any radiation detected will consist of only gamma rays.

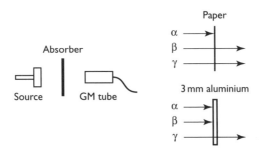

Worked example

A student carried out an investigation to determine how the thickness of lead affected the absorption of gamma rays.

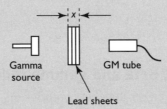

Lead discs were placed between the source and the detector, and for each value of thickness, x, the count rate, N, was measured.

The student's results are given in the table below:

x/mm	0	2.0	4.0	6.0	8.0	10
N/s^{-1}	880	665	503	380	287	217

(a) Plot a graph of the count rate against thickness.
(b) Use the graph to determine the thickness of lead needed to reduce the count to one half of its original value (the half-thickness of lead).
(c) The absorption of gamma rays can be represented by the expression, $N = N_0 e^{-\mu x}$, where μ is the absorption coefficient of the material. Use the student's results to plot a suitable graph to obtain a value of the absorption coefficient of lead.
(d) A fellow student stated that she had been told to subtract the value of the background radiation from all radiation measurements. How would such a course of action have affected the student's results?

Answer

(a)

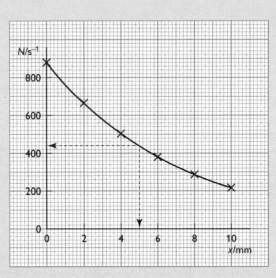

(b) From the graph, the thickness when $N = 440\,s^{-1}$ is 5.0 mm.

(c) Taking the natural log of each side of the equation $N = N_0 e^{-\mu x}$ gives
$\ln N = \ln N_0 - \mu x$.

A graph of $\ln N$ against x will give a straight line of gradient $-\mu$ and intercept $\ln N_0$.

x/mm	0	2.0	4.0	6.0	8.0	10
N/s⁻¹	880	665	503	380	287	217
ln(N/s⁻¹)	6.79	6.50	6.22	5.94	5.66	5.38

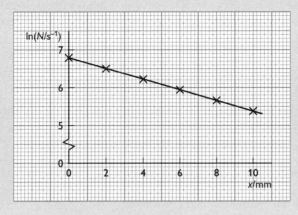

From the graph: gradient $= \dfrac{(5.38 - 6.79)}{10 \times 10^{-3}\,\text{m}}$

$= -140\,\text{m}^{-1}$

and so $\mu = 140\,\text{m}^{-1}$

Note: The expression $N = N_0 e^{-\mu x}$ is not given in the specification, and you are not expected to know it. The mathematical requirements include the use of logarithms to determine constants in exponential decays, and you may meet an unfamiliar equation that relates to the topic in the examination.

(d) The background radiation is generally about 30 counts per minute, and so, in this case, it is too small to affect the results.

Stability

Earlier in this section, an $N - Z$ graph for stable nuclei was drawn, and the alpha and beta decay processes were discussed. An unstable nucleus is one in which the ratio of neutrons to protons is such that there is a probability that an alpha particle or beta particle is emitted. Some nuclei are less stable than others and may decay in fractions of a second. The stability is represented by the **decay constant**, λ, which is defined from the equation:

$$\frac{dN}{dt} = -\lambda N$$

This means that for a sample of N nuclei, the rate of decay (i.e. the number of disintegrations per second) depends on the number of nuclei in the sample and the decay constant.

It must be stressed that such probability-governed decays occur **spontaneously** and **randomly**. This means that a nucleus has no precise lifetime, and may decay at any time. The probability of decay can be likened to the probability of dice landing with the six-side up — if 600 dice are thrown, the probability is that 100 will show a six, but it is unlikely to occur every time. Even minute quantities of radioactive material contain vast numbers of nuclei, so statistical analysis can be applied to radioactive decays.

Those of you studying mathematics will recognise the above expression as an example of a differential equation, a solution of which is $N = N_0 e^{-\lambda t}$, where N_0 represents the number of nuclei when $t = 0\,\text{s}$.

Both equations above are on the exam data sheet.

Half-life

Because radioactive decays are exponential, with the nuclei falling by a fixed fraction in a given time, the lifetime of a substance is difficult to ascertain. To measure the stability, the half-life is used:

• **Half-life**, $t_{\frac{1}{2}}$, is the average time taken for the activity of a radioactive material to reduce to one-half of its original value.

The half-life of a short-lived isotope can be measured in the laboratory using a suitable detector to measure values of the activity of an isolated sample of the substance. A graph of activity against time is plotted (or produced from a data logger) and the half-life determined from the curve.

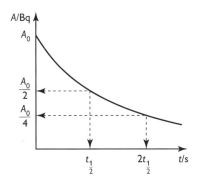

The half-life is related to the decay constant. If an isotope has a large decay constant then it will decay quickly, and so have a short half-life. From the expressions for radioactive decay used above, it is possible to derive a relationship between half-life and the decay constant:

$$\lambda = \frac{\ln 2}{t_{\frac{1}{2}}}$$

This expression is also included in the data sheet.

Worked example

(a) The activity of a sample of radon-220 gas is measured as 480 Bq. After a time of 2 minutes 48 seconds, the reading has fallen to 60 Bq. Calculate the half-life of the radon.

(b) A sample of rock contains 0.5 g of uranium-238, and has an activity of 6.3×10^3 Bq. If 1 g of uranium contains 2.6×10^{21} atoms, calculate the half-life of uranium-238.

(c) Protactinium-234 decays to uranium-234 with a half-life of 72 s. How long will it take for 90% of the protactinium to decay?

Answer

(a) For the activity to fall from 480 Bq to 60 Bq it takes three half-lives
$$480 \rightarrow 240 \rightarrow 120 \rightarrow 60$$
$3 \times$ half-lives $= 168$ s
So the half-life is 56 s

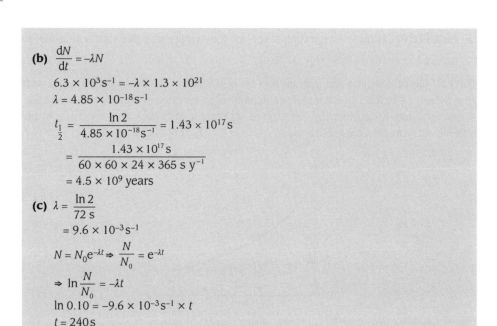

(b) $\dfrac{dN}{dt} = -\lambda N$

$6.3 \times 10^3 \, s^{-1} = -\lambda \times 1.3 \times 10^{21}$

$\lambda = 4.85 \times 10^{-18} \, s^{-1}$

$t_{\frac{1}{2}} = \dfrac{\ln 2}{4.85 \times 10^{-18} \, s^{-1}} = 1.43 \times 10^{17} \, s$

$\qquad = \dfrac{1.43 \times 10^{17} \, s}{60 \times 60 \times 24 \times 365 \, s \, y^{-1}}$

$\qquad = 4.5 \times 10^9$ years

(c) $\lambda = \dfrac{\ln 2}{72 \, s}$

$\qquad = 9.6 \times 10^{-3} \, s^{-1}$

$N = N_0 e^{-\lambda t} \Rightarrow \dfrac{N}{N_0} = e^{-\lambda t}$

$\Rightarrow \ln \dfrac{N}{N_0} = -\lambda t$

$\ln 0.10 = -9.6 \times 10^{-3} \, s^{-1} \times t$

$t = 240 \, s$

Applications of radioactive materials

There are numerous applications of radioactive materials. A few common uses are given in this section, but you should be aware that different situations may be examined. However, the basic principles will be the same.

Radioactive dating provides a method of determining the age of rocks, fossils and organic material. The age of the Earth can be estimated by comparing the amounts of uranium-238 with the stable decay product lead-206 present in rocks. An approximate 50:50 mix tells us that the Earth has existed for about one half-life of the uranium; that's around 4500 million years. Carbon dating compares the percentage of the radioactive isotope carbon-14 in ancient organic materials with that in similar living organisms. Carbon-14 decays by beta-minus emission to nitrogen with a half-life of 5730 years. When respiration or photosynthesis ceases in an organism, the carbon-14 is not replenished so the levels fall and the age can be estimated.

Smoke alarms utilise the ionising properties of alpha particles. A small americium-241 source emits alpha particles into the space between two electrodes with a 9 volt potential difference across them. The particles ionise the air so that a small current flows between the electrodes. When smoke enters the space, the ionisation is suppressed by the larger smoke particles and this triggers the alarm. Because the alpha particles are absorbed by the casing and have such a small range, they present a very low-level health hazard.

Radioactive tracers are used to follow the progress of fluids. Oil and gas companies can check for leaks in pipelines by measuring the activity of the tracer at each end of a pipe. In medicine, iodine-123 is taken up by the thyroid gland, and external readings of the emissions from the gland can be used to measure the metabolic rate of the organ.

Medical imaging uses compounds containing a pure gamma-emitter, such as technetium-99m, that are absorbed by specific organs. The gamma radiation is emitted from the organ and leaves the body with very little absorption by the tissue — it is detected using a gamma camera. The image enables the physician to detect 'hot spots' due to tumours, or blockages within the organ.

Radiotherapy uses high-energy gamma rays from a cobalt-60 source focused on a tumour. The rays kill the malignant cells. An alternative method of cancer treatment is to target the affected region with powerful beta-emitters.

In all the above applications, and in others not described, it is important that the type of radiation, its strength and the half-life of the isotope are appropriate for the intended purpose. For example, it would be futile to use an alpha source for radioactive imaging because all the radiation would be absorbed by the body.

Oscillations

In Unit 2 you studied wave motion, and learned some definitions relating to the oscillations of particles within the wave. In Unit 4, the period of an oscillation was linked to the angular velocity of a rotating point. These definitions and relationships will be used extensively in this section and so they are repeated below:

- the **amplitude**, A, is the maximum displacement of a particle about the equilibrium position
- the **frequency**, f, is the number of oscillations completed every second
- the **period**, T, is the time taken for one complete oscillation
- $T = \dfrac{1}{f}$
- $T = \dfrac{2\pi}{\omega}$

Simple harmonic motion

Simple harmonic motion (SHM) is a particular form of oscillatory motion common to most vibrating systems. It can be described in terms of a mass held in equilibrium between two stretched springs.

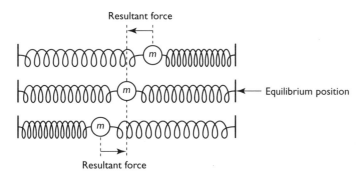

If the mass is displaced to the right, the tension in the left-hand spring is increased, while that in the right-hand spring is reduced. The mass now has a resultant force towards the equilibrium position, and when released will accelerate towards this point. If the displacement is increased, the resultant force will be bigger, resulting in a greater acceleration. Similarly, if the displacement is to the left, there will be a resultant force to the right — i.e. back towards the equilibrium position. If the mass is displaced and then released, it will oscillate about the equilibrium position with simple harmonic motion.

- **Simple harmonic motion** is the oscillatory motion about an equilibrium position such that the acceleration is proportional to the displacement, and always directed towards the equilibrium position.

Mathematically, this is expressed as:

$$a = -\omega^2 x$$
$$\text{or } a = -\left(\frac{2\pi}{T}\right)^2 x \text{ or } a = -(2\pi f)^2 x$$

It should be noted that displacement and acceleration are *vector* quantities, so if a displacement is taken as positive to the right, the acceleration will have a negative value — i.e. to the left.

The graph of the equation $a = -\omega^2 x$ is a straight line of gradient $-\omega^2$, with the line extending as far as the amplitude, A, in each direction.

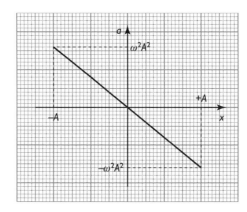

Worked example

The graph shows how the acceleration of a body varies with its displacement.

Use the graph to determine T, the period of oscillation of the body.

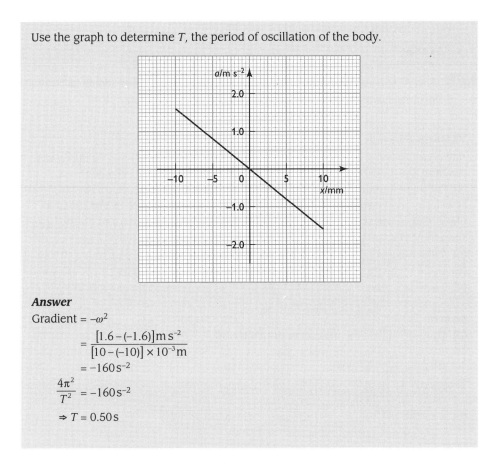

Answer

Gradient $= -\omega^2$

$$= \frac{[1.6-(-1.6)]\,\mathrm{m\,s^{-2}}}{[10-(-10)]\times 10^{-3}\,\mathrm{m}}$$

$$= -160\,\mathrm{s^{-2}}$$

$$\frac{4\pi^2}{T^2} = -160\,\mathrm{s^{-2}}$$

$$\Rightarrow T = 0.50\,\mathrm{s}$$

Displacement–time graph for SHM

If the transmitter of a displacement sensor is attached to the bottom of a mass on a spring, when the mass is pulled down and then released its subsequent motion can be tracked using the sensor interfaced with a data logger.

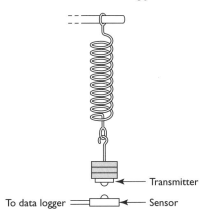

It can be shown that the resulting graph has the form:

$x = A\cos\omega t$

Tip Always ensure that your calculator is in radian mode when doing calculations involving functions of ω.

Worked example

A mass on a spring is displaced 0.036 m vertically downwards from its equilibrium position. It is then released. A stopwatch is started when the mass reaches its highest point and the oscillations are counted. The mass takes 7.60 s to perform 20 oscillations. Assuming the motion is SHM, it can be described by the equation $x = A\cos 2\pi ft$, where x is the displacement in the upward direction and t is the time since the clock was started.

(a) What are the values of **(i)** A and **(ii)** f for this motion?

(b) Use the equation to calculate the displacement when $t = 0.50$ s.

(c) Sketch a graph of displacement against time for two complete oscillations.

Answer

(a) (i) Amplitude is the maximum displacement, $A = 0.036$ m

(ii) $T = \dfrac{7.60\,\text{s}}{20} = 0.38\,\text{s}$

$f = \dfrac{1}{T} = \dfrac{1}{0.38\,\text{s}} = 2.6\,\text{Hz}$

(b) $x = A\cos 2\pi ft$

$= 0.036\,\text{m} \times \cos(2\pi\,\text{rad} \times 2.6\,\text{s}^{-1} \times 0.50\,\text{s})$

$= -0.011\,\text{m}$

Displacement $= -0.011$ m; the spring is 11 mm *below* its equilibrium position.

(c)

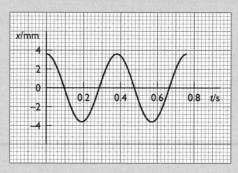

Note: A 'sketch' graph should still be drawn reasonably accurately, showing the amplitude as 36 mm and the period as 0.38 s. You can check that the answer for part **(b)** is correct by reading off the displacement at 0.50 s. As this is a sketch graph, your answer will only be approximate.

content guidance

Velocity–time and acceleration–time graphs for SHM

The velocity of a body undergoing SHM can be found by working out the gradient of its displacement–time graph. It is clear that the gradient is zero at the extremes of the motion, and that the velocity will be a positive or negative maximum at the equilibrium position.

The velocity can be represented by the expression:

$v = A\omega\sin\omega t$

Similarly, the acceleration can be found using the gradient of a velocity–time graph, and can be represented, in SHM, by the expression:

$a = -A\omega^2\cos\omega t$

These graphs are best illustrated by drawing one directly below the other:

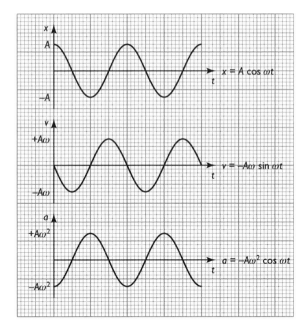

The velocity–time graph in the diagram above is strictly a negative sine graph, but this represents the direction of the motion.

The graphs show that the maximum values of velocity occur at the equilibrium point, and those of acceleration at each end of the oscillation. The maximum values of both $\sin\omega t$ and $\cos\omega t$ are ± 1, so those of velocity and acceleration are:

$v_{max} = \pm A\omega$

$a_{max} = \pm A\omega^2$

Worked example

A data logger is used to display graphs of displacement and velocity against time for the oscillations of a mass on a spring. The diagram shows the displacement–time graph.

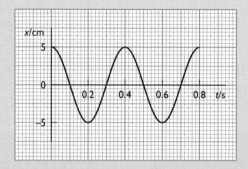

(a) Use the graph to determine **(i)** the amplitude, **(ii)** the frequency, **(iii)** the maximum velocity of the mass.

(b) Sketch the velocity–time graph for the oscillation.

Answer

(a) (i) The amplitude is the maximum displacement from the equilibrium position, and this is 0.05 m.

(ii) The period is the time for one oscillation, and this is 0.4 s.

$$f = \frac{1}{T}$$
$$= \frac{1}{0.4\,s}$$
$$= 2.5\,Hz$$

(iii) This can be answered in one of two ways. You can use the values of A and T in the equation $v_{max} = \pm A\omega = \pm A2\pi f$

$$v_{max} = \pm 0.05\,m \times 2 \times \pi \times 2.5\,Hz$$
$$= 0.79\,m\,s^{-1}$$

Or you can determine the gradient of the displacement-time graph as it passes through the equilibrium position. This is achieved by drawing a tangent to the line as it crosses the $x = 0$ axis. This should give a value of around $0.8\,m\,s^{-1}$, but it is quite difficult to draw tangents accurately, so the first method is preferable.

(b) As in the previous worked example, the 'sketch' graph should be drawn carefully, starting at the origin, with the correct period and the amplitudes shown.

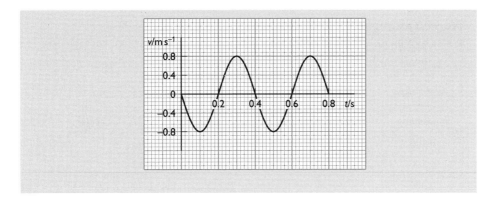

Mass–spring system

One of the more common types of oscillator is a mass on a spring. Oscillating atoms can be modelled on similar systems.

In Unit 1 you looked at Hooke's law for extending springs and wires. An expression for the relationship between the applied force and the extension of the spring was deduced:

$F = -kx$, where k is the stiffness or the spring constant.

For an oscillating mass on a spring, this restoring force is responsible for the acceleration of the displaced mass back towards the equilibrium position. By applying Newton's second law, $F = ma$, to the oscillating system:

$F = -m\omega^2 x$

Combining the two equations:

$$m\omega^2 = k$$
$$\Rightarrow \omega^2 = \frac{k}{m}$$
$$\Rightarrow \omega = \sqrt{\frac{k}{m}} = \frac{2\pi}{T}$$

This leads to the mechanical oscillator equation:

$$T = 2\pi\sqrt{\frac{m}{k}}$$

This formula tells us a good deal about mechanical oscillators. The period for a large mass on weak spring will be long (imagine the slow oscillations of an elephant on a bungee jump). Small masses with stiff elastic connections (e.g. atoms in a solid) will oscillate at very high frequencies.

Worked example

In an experiment to measure the stiffness of a spring, weights were added to the spring and the extension was noted for each weight. For each load, the mass was made to oscillate, and the time for 20 oscillations was measured. A set of readings is given below:

Mass m/kg	Weight W/N	Extension x/mm	$20T$/s	T^2/s^2
0.100		8	3.60	
0.200		16	5.10	
0.300		25	6.22	
0.400		31	7.18	
0.500		41	8.03	
0.600		48	8.80	

(a) Complete the table to include the values of W and T^2.
(b) Plot W against x. Use the graph to determine a value for the spring constant, k.
(c) Plot a graph of T^2 against m. Use your graph to determine the spring constant, k.

Answer

(a)

Mass m/kg	Weight W/N	Extension x/mm	$20T$/s	T^2/s^2
0.100	0.98	8	3.60	0.032
0.200	1.96	16	5.10	0.065
0.300	2.94	25	6.22	0.097
0.400	3.92	31	7.18	0.129
0.500	4.90	41	8.03	0.161
0.600	5.88	48	8.80	0.193

(b)

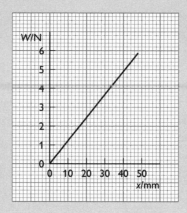

$$k = \text{gradient}$$
$$= \frac{5.88\,N}{48 \times 10^{-3}\,m}$$
$$= 120\,N\,m^{-1}$$

(c)

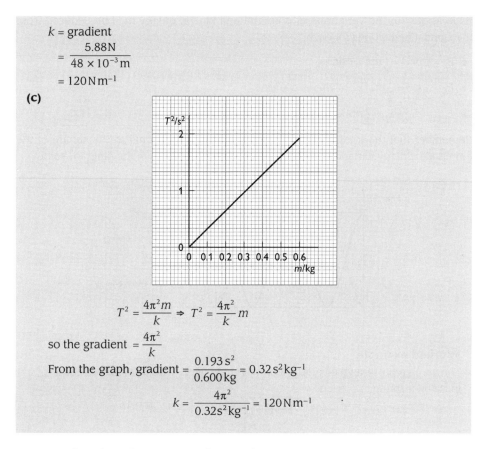

$$T^2 = \frac{4\pi^2 m}{k} \Rightarrow T^2 = \frac{4\pi^2}{k}\,m$$

so the gradient $= \dfrac{4\pi^2}{k}$

From the graph, gradient $= \dfrac{0.193\,s^2}{0.600\,kg} = 0.32\,s^2\,kg^{-1}$

$$k = \frac{4\pi^2}{0.32\,s^2\,kg^{-1}} = 120\,N\,m^{-1}$$

Energy in simple harmonic oscillators

You will be familiar with the energy transfers during the cycle of a simple pendulum. When the bob is raised to the extreme position, it has a maximum value of gravitational potential energy. At the centre of the swing, the bob is at the position of minimum gravitational potential energy, and the loss in potential energy has been transferred as kinetic energy. The pendulum motion is approximately simple harmonic for small oscillations, and the transfer of potential and kinetic energy is typical of all simple harmonic oscillators.

The model of the mass between two springs can be used to illustrate this energy transfer. When the mass is displaced from the equilibrium position, elastic potential energy is transferred to the spring. On release, this is converted to kinetic energy as the mass accelerates until, at the mid-point, the mass has maximum velocity and kinetic energy.

If no energy is lost from the system, the energy of an oscillating mass will continuously interchange between potential and kinetic, with maximum values of potential energy at the extreme points and maximum kinetic energy at the centre.

Expressions for the variations of potential and kinetic energy may be deduced from the standard equations for each:

- $E_k = \frac{1}{2}mv^2 = \frac{1}{2}mA^2\omega^2\sin^2\omega t$
- $E_p = \frac{1}{2}kx^2 = \frac{1}{2}m\omega^2x^2 = \frac{1}{2}mA^2\omega^2\cos^2\omega t$
- $E_{total} = E_k + E_p = \frac{1}{2}mA^2\omega^2(\sin^2\omega t + \cos^2\omega t) = \frac{1}{2}mA^2\omega^2$

Remember that the total energy in an oscillator is proportional to the square of the amplitude. Energy variations with time can be represented as in the diagram below:

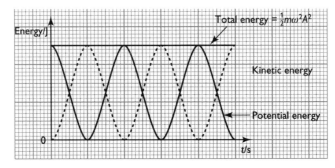

Worked example

The graph represents the variation of potential energy with displacement of a mass on a spring.

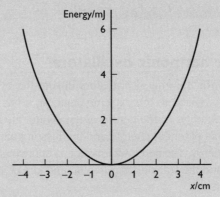

(a) Sketch the graph. Add labelled lines to show the variation with displacement of (i) the kinetic energy of the mass, and (ii) the total energy in the system.

> **Tip** Always check that the velocity is maximum at the centre, and zero at the ends for the kinetic energy curve; and zero at the centre, and maximum at the ends for the potential energy curve.

(b) Use the graph to determine the value of the constant k, the stiffness of the spring.

Answer

(a)

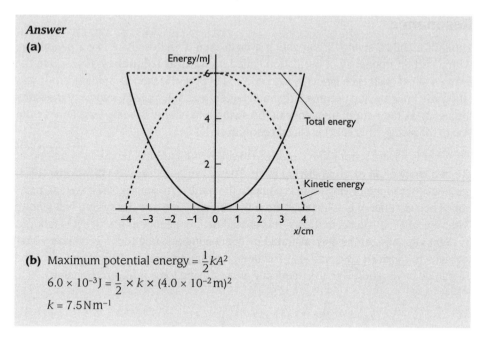

(b) Maximum potential energy $= \frac{1}{2}kA^2$

$6.0 \times 10^{-3}J = \frac{1}{2} \times k \times (4.0 \times 10^{-2}m)^2$

$k = 7.5\,N\,m^{-1}$

Free and forced oscillations

If no energy loss occurred from an oscillator, the motion would continue indefinitely. Such oscillations are said to be **free oscillations**. In practice, energy is transferred to the surroundings by viscous forces (e.g. air resistance) or internally to the system (e.g. vibrations giving rise to thermal energy or sound). The motion is said to be **damped**, resulting in a decrease in the energy (and hence amplitude) of the oscillations.

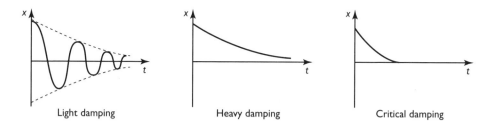

If a system is *lightly* damped, the amplitude decreases approximately exponentially with time — an example is a pendulum oscillating in air. When the damping is *heavy*, the system returns slowly to its equilibrium position — imagine a spring trying to oscillate in thick oil. When the time taken for the displacement to become zero is a minimum, the system is said to be *critically* damped — as with, for example, the shock absorbers of a car. If oscillations are forced upon ductile materials, energy will be absorbed due to plastic deformation, and the motion will be damped. This will be noticeable in overstretched springs, and is useful in sound insulation.

Resonance

Think of a child's swing. If you give it a push then it will oscillate like a pendulum, with a certain frequency of oscillation called its **natural frequency**. If you give the swing a small push *each time it reaches the end of its swing* then its amplitude of oscillation will build up and become larger and larger. You are applying a force at the same frequency as the natural frequency of the swing and energy is being transferred from you to the swing. This effect is called **resonance**.

All systems have a natural frequency of vibration, but can also be made to vibrate at the frequency of an external driving force. These are called **forced vibrations**. When the frequency of the driving force is equal to the natural frequency of the system, resonance occurs. Energy is transferred from the driving force and the system will vibrate with *very large amplitude*. In theory, this should be infinite, but in practice it is reduced by damping. This can be demonstrated by connecting a spring to a mechanical vibrator, which is driven by a variable frequency signal generator. The variation of the amplitude of vibration with the frequency of the driving force is illustrated below.

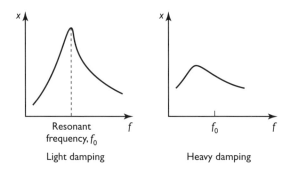

Light damping — Resonant frequency, f_0 Heavy damping

The second graph shows that if the damping is increased then the amplitude is reduced, and resonance occurs at a lower frequency.

There are many instances where resonance occurs. Some examples are:

- **Sound** — an opera singer can produce a pure note with a frequency equal to the natural frequency of a brandy glass, and the resulting large amplitude oscillations cause the glass to shatter.
- **Microwaves** — the frequency of the microwaves in an oven is equal to the natural frequency of oscillation of the polar water molecules in food, so energy is readily transferred, increasing the internal energy and, hence, the temperature of the food.
- **Radios** — when the signal received by an aerial has the same frequency as the natural frequency of the oscillator in the tuning circuit, the large amplitude of the oscillating electrons enables the selection of particular individual stations.

Astrophysics and cosmology

This topic covers the physical interpretation of astronomical observations, the formation and evolution of the stars, and the history and future of the universe. By necessity, the specification covers a limited section of the topic, but wider reading will give you a greater understanding of the basic concepts needed for the examination.

Gravitational fields

In Unit 4 you studied the ideas of electric and magnetic fields. You should recall that a **field** is a region in which a force is experienced — for example, a particle with charge q in an electric field of strength E, will be subject to a force of Eq. In a gravitational field, objects with mass will experience a gravitational force:

$$g = \frac{F}{m} \qquad \text{unit: N kg}^{-1}$$

Newton's law of gravitation relates the force between two point masses in terms of their magnitude and their separation. It can be represented by the expression:

$$F = \frac{Gm_1 m_2}{r^2}$$

G is the universal gravitational constant; its value is $6.67 \times 10^{-11}\,\text{N m}^2\,\text{kg}^{-2}$.

Combining these two expressions, an equation for the gravitational field strength at a distance r from a point mass m can be deduced:

$$g = \frac{Gm}{r}$$

For extended masses, the distance between the centres of mass can be used. For example, the gravitational force between the Earth and a mass m on its surface equals $\frac{GMm}{R^2}$ where M is the mass of the Earth and R is its radius. So the gravitational field strength, g_0, at the Earth's surface is:

$$g_0 = \frac{GM}{R^2} = 9.8\,\text{N kg}^{-1}$$

Worked example
(a) Show that the mass of the Earth is about $6.0 \times 10^{24}\,\text{kg}$, assuming that its mean radius is 6370 km.
(b) Calculate the gravitational force acting on the International Space Station. The mass of the station is about $3.00 \times 10^5\,\text{kg}$ and the station orbits the Earth at an altitude of about 350 km.
(c) The Moon has a mean radius of 1740 km and the gravitational field strength on its surface is about one-sixth of that of the Earth. Show that the mean density of the Moon is about 60% of the Earth's density.

Answer

(a) $g_0 = \dfrac{GM}{R^2} \Rightarrow M = \dfrac{g_0 R^2}{G}$

$M = \dfrac{9.8 \text{ N kg}^{-1} \times (6370 \times 10^3 \text{m})^2}{6.67 \times 10^{-11} \text{N m}^2 \text{ kg}^{-2}}$

$= 5.96 \times 10^{24} \text{ kg}$

> **Tip** Give your answer to one more significant figure in 'show that' questions.

(b) $F = \dfrac{GMm}{r^2} = \dfrac{GMm}{(R+h)^2}$

$= \dfrac{6.67 \times 10^{-11} \text{N m}^2 \text{ kg}^{-2} \times 5.96 \times 10^{24} \text{kg} \times 3.00 \times 10^5 \text{ kg}}{(6370 \times 10^3 \text{m} + 350 \times 10^3 \text{m})^2}$

$= 2.6 \times 10^6 \text{ N}$

(c) To obtain an expression for field strength in terms of density, use the relationship:

$m = V \times \rho = \dfrac{4}{3}\pi r^3 \rho$

$\Rightarrow g = \dfrac{G \dfrac{4}{3}\pi r^3 \rho}{r^2} = G\tfrac{4}{3}\pi r \rho$

$\dfrac{g_m}{g_e} = \dfrac{G \dfrac{4}{3}\pi r_m \rho_m}{G \dfrac{4}{3}\pi r_e \rho_e}$

$= \dfrac{r_m \rho_m}{r_e \rho_e} = \dfrac{1}{6}$

$\dfrac{\rho_m}{\rho_e} = \dfrac{g_m r_e}{g_e r_m}$

$= \dfrac{1}{6} \times \dfrac{6.37 \times 10^6 \text{m}}{1.74 \times 10^6 \text{m}}$

$= 0.61 \text{ or } 61\%$

Satellite and planetary motion

In Unit 4 you studied circular motion and learned that a centripetal force is required for a mass to move with constant speed in a circle:

$F = \dfrac{mv^2}{r}$

or $F = m\omega^2 r = m\dfrac{4\pi^2}{T^2}r$

For gravitational forces:

$\dfrac{GMm}{r^2} = \dfrac{mv^2}{r} = m\dfrac{4\pi^2}{T^2}r$

which leads to $\boldsymbol{v^2 = \dfrac{GM}{r}}$ and $\boldsymbol{T^2 = \dfrac{4\pi^2}{GM}r^3}$

Worked example

(a) Calculate the speed of the International Space Station as it orbits the Earth at a height of 350 km (mass of Earth = 6.0×10^{24} kg; radius of Earth = 6400 km).

(b) Calculate the altitude of a communications satellite in a geostationary orbit above the Equator.

Answer

(a) $v^2 = \dfrac{GM}{(R+h)}$

$$= \frac{6.67 \times 10^{-11} \, \text{N} \, \text{m}^2 \, \text{kg}^{-2} \times 6.0 \times 10^{24} \, \text{kg}}{(6400 \times 10^3 + 350 \times 10^3) \, \text{m}}$$

$= 5.93 \times 10^7 \, \text{m}^2 \, \text{s}^{-2}$

$v = 7700 \, \text{m} \, \text{s}^{-1}$

(b) To remain in a geostationary orbit, the satellite must have a period of 24 hours.

$$T^2 = \frac{4\pi^2}{GM} r^3$$

$$r^3 = \frac{GMT^2}{4\pi^2}$$

$$= \frac{6.67 \times 10^{-11} \, \text{N} \, \text{m}^2 \, \text{kg}^{-2} \times 6.0 \times 10^{24} \, \text{kg} \times (24 \times 60 \times 60 \, \text{s})^2}{4\pi^2}$$

$\Rightarrow r = 4.23 \times 10^7 \, \text{m} = 42\,300 \, \text{km}$

Altitude = $(r - R)$ = (42 300 km – 6400 km) = 36 000 km

Gravitational and electric fields

In many respects, gravitational and electric fields are similar. Newton's law $F = \dfrac{Gm_1 m_2}{r^2}$ and Coulomb's law ($F = \dfrac{kQ_1 Q_2}{r^2}$) are both inverse square laws representing the forces between point masses and point charges, respectively. The major difference is that gravitational forces are always attractive, whereas electric forces can be attractive or repulsive. Gravitational and electric field comparisons are listed in the table below.

Gravitational field	Electric field
Acts on mass	Acts on charge
g = force per unit mass	E = force per unit charge
Unit: N kg^{-1}	Unit: N C^{-1}
Field strength is proportional to mass	Field strength is proportional to charge
Field obeys inverse square law	Field obeys inverse square law
Results in attraction only	Can result in attraction or repulsion

Like electric fields, gravitational fields can be represented by field lines:

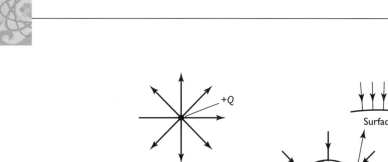

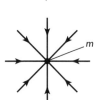

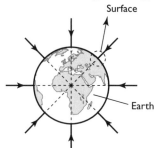

From the diagram it can be seen that, for an extended mass like the Earth, the field lines close to the surface are almost parallel, and so the field may be considered to be uniform.

Stars

Stars are massive bodies that emit vast amounts of radiant energy from nuclear reactions within them. On a clear night we can observe hundreds of stars with the naked eye, but millions more exist in the universe. The observed brightness varies considerably and will depend on factors such as the size, surface temperature and distance of a star from the observer. In this section you will study these factors and how they are related, together with the changes that take place during the life cycle of a star.

There are two significant definitions relating to the radiation from a star. They are:

- **Luminosity**, L, the total power emitted by the star — the unit is the watt, W
- **Radiant flux** (intensity), F, the power per unit area received by an observer — the unit is $W\,m^{-2}$

It is very important that you understand the difference between these. Luminosity relates to the power *emitted* by a star; the flux relates to that *received* at some point (usually on Earth). The two quantities are related in terms of the distance, d, between the star and the observer by the expression:

$$F = \frac{L}{4\pi d^2}$$

Worked example

The radiant flux from the Sun is measured as $1.38\,kW\,m^{-2}$ at the upper edge of the Earth's atmosphere.

(a) If the mean radius of the Earth's orbit is $1.49 \times 10^{11}\,m$, calculate the luminosity of the Sun.

(b) Why is the flux measurement taken above the atmosphere?

Answer

(a) $F = \dfrac{L}{4\pi d^2} \Rightarrow L = F\,4\pi d^2$

$L = 1.38 \times 10^3\,\text{W}\,\text{m}^{-2} \times 4\pi \times (1.49 \times 10^{11}\,\text{m})^2$

$\quad = 3.85 \times 10^{26}\,\text{W}$

(b) The atmosphere absorbs some of the Sun's radiation.

Measuring the distances of stars from Earth

The distances of stars from Earth are usually found by comparing the flux measured for the unknown star and comparing it with a star of the same type whose luminosity is known. Such a star is called a **standard candle**. If the luminosity is known and the flux is measured, the distance can be calculated using the inverse square law relationship.

Stellar distances are extremely large and so are usually given in **light years** — the distance travelled in 1 year when moving at the speed of light.

Worked example

The luminosity of Vega is $1.93 \times 10^{28}\,\text{W}$. Calculate the distance of Vega from Earth if the radiant flux at the edge of the Earth's atmosphere is $1.30 \times 10^{-8}\,\text{W}\,\text{m}^{-2}$. Convert your answer to light years.

Answer

$F = \dfrac{L}{4\pi d^2} \Rightarrow d = \sqrt{\dfrac{L}{4\pi F}}$

$d = \sqrt{\dfrac{1.93 \times 10^{28}\,\text{W}}{4\pi \times 1.30 \times 10^{-8}\,\text{W}\,\text{m}^{-2}}}$

$\quad = 3.44 \times 10^{17}\,\text{m}$

To convert into light years:

$d = \dfrac{3.44 \times 10^{17}\,\text{m}}{3.00 \times 10^8\,\text{m}\,\text{s}^{-1} \times (365.25 \times 24 \times 60 \times 60)\,\text{s}}$

$\quad = 36.3\text{ light years}$

Parallax method

If you move your head from side to side when looking out of a window, you will notice that close objects, for example the window frame, move relative to the more distant background. This effect is known as **parallax**, and it provides a method of determining the distance of stars that are relatively close to our solar system.

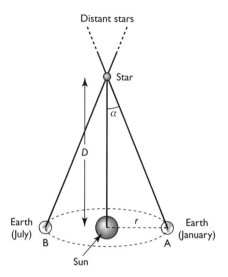

If the position of the star against the very distant background stars is noted on 1 January to be position A, and then again on 1 July to be position B, it appears to have moved through an angle of 2α (much smaller than shown on the diagram — usually less than 1 second of arc or $\frac{1}{3600}$ of a degree). For such small angles $\alpha = \frac{r}{D}$ radians, where r is the radius of the Earth's orbit around the Sun. Astronomers use the unit **parsec**, which is the distance of a star with a parallax angle of 1 second of arc:

- 1 parsec (pc) = 3.09×10^{16} m = 3.26 light years

Note: You will not be expected to know the definition or value of the parsec, but you may see measurements in parsecs with the conversion rate given.

It must be emphasised that the parallax method is only used for nearby stars, because the angles for distant bodies become too small to be measured accurately.

Using Cepheid variables

For stars in distant galaxies, Cepheid variable stars are used as standard candles. The luminosity of a Cepheid is related to the period at which it varies, so the luminosity can be found by measuring the period. The distance is estimated using the measured flux incident on the Earth and the inverse square law as before.

Life cycle of stars

A Hertzsprung–Russell (HR) diagram is a chart showing the luminosity and temperature of a large number of known stars at various stages of their life cycles. It is important to note that the scales are logarithmic in order to cover the large variations in luminosity and temperature, and that the temperature scale increases to the *left*. The life cycle of a star similar to our Sun is shown by the line ABCD on the diagram below.

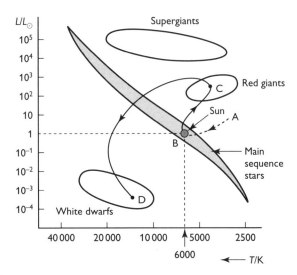

Star formation

Stars are formed from interstellar cloud, where there is sufficient material (predominately hydrogen) at a high enough density, to cause a gravitational collapse towards the centre of mass of the cloud. The gravitational potential energy of the particles is converted to kinetic energy as they rush inwards, so the gas becomes more dense and its internal energy, and hence the temperature, rise greatly (point A on the chart). When the temperature reaches about $8 \times 10^6 \, K$, nuclear fusion begins (see later in this section) and vast amounts of energy are released in the core of the star. The core has a radius of about a quarter of the total radius of the star and is surrounded by the envelope of hydrogen in which no fusion takes place.

Main sequence

A star spends most of its life as a main sequence star. The star has reached equilibrium between the gravitational forces and the forces due to the radiation pressure from the fusion of hydrogen to form helium in the core. A main sequence star similar to our Sun has a lifetime of about 10^{10} years, during which time its size, mass and luminosity remain fairly constant. More massive stars fuse hydrogen at a much greater rate and, despite having more fuel, have a shorter lifetime.

Red giants

When all the hydrogen in the core has been converted to helium, there is no outward radiation pressure, and the core collapses due to gravitational forces. The thermal energy produced increases the temperature to about $10^8 \, K$, at which point helium fusion begins in a much diminished core. Surrounding this core is a shell of hydrogen which is heated sufficiently to initiate fusion, and produce an outward pressure which prevents the star from collapsing and makes it expand to several hundred times its former size. The outer regions consist of low-density hydrogen that has cooled to give a surface temperature of 3000–4000 K, making it appear red. Because the surface area is so much larger than that of the main sequence star, the luminosity is much bigger. The outer layers will drift into space and, after a life of about 10^8 years, the helium in the core will have fused into carbon and oxygen, and the outer envelope is ejected into space in the form of cooler, thinner matter known as a **planetary nebula**.

White dwarfs

When the helium-burning ceases, the temperature is not high enough for any further fusion. When all the outer shells have blown away, the high temperature core of carbon 'ash' remains. Because this is so small (about the size of the Earth) its luminosity falls, but its temperature is high enough to make it 'white hot'. No further fusion occurs within a white dwarf and it continues to cool down. After a few thousand million years, its surface temperature will approach absolute zero, ending its life as a burnt-out cinder, sometimes called a **black dwarf**.

This cycle for a 'Sun-like' star is represented by the line ABCD on the Hertzsprung–Russell diagram. You must be aware that this line represents the variations of luminosity and temperature and *not* the change in position of the star.

Larger stars

The cycle shown on the HR diagram is for stars of similar mass to the Sun. If a star has a mass greater than three times that of the Sun it will behave differently when it leaves the main sequence. After the helium has completely fused, the core contracts as before, but now there is sufficient mass to trigger the fusion of heavier elements. The outer layers are pushed outwards at each stage and **supergiant** stars are formed.

Fusion in the core continues until iron is produced. Elements of higher mass than iron will no longer produce energy by fusion, so, no matter how big the star is, the core will collapse dramatically. Protons are forced to combine with electrons to produce neutrons, which combine together to produce a massively dense sphere. The collapse of the outer shells of the star causes the in-falling material to bounce off the rigid core, and the shock wave produced throws out the outer layers with tremendous energy. This gigantic explosion is called a **supernova** and can produce enough energy to temporarily outshine the whole galaxy. Unlike the demise of the smaller stars that take place over millions of years, a supernova explosion is very rapid. The core remnant exists as a **neutron star** with a radius of several kilometres and a density of about 10^{18} kg m^{-3}.

If a main sequence star has a mass greater than about 10 solar masses, the gravitational forces on the collapsing core are so great that that its size approaches zero and its density becomes virtually infinite. Such a singularity is termed a **black hole** with a gravitational field so powerful that not even light can escape from it.

Black body radiators

A black body is one that will absorb all wavelengths of radiation falling on it, at all temperatures. It is a strange fact that black bodies are also perfect emitters of radiation. The Sun and other stars behave as black bodies — if you were to shine light on them, none would be reflected back. This means that the energy flux emitted depends only on the body's temperature and not its composition.

There are two laws governing the emission of radiation from a black body:

- **Stefan–Boltzmann law:** L = surface area × σT^4 = $4\pi r^2 \sigma T^4$

  σ is the Stefan–Boltzmann constant = $5.67 \times 10^{-8}\,\mathrm{W\,m^{-2}\,K^{-4}}$
- **Wien's law:** $\lambda_{max}T = 2.898 \times 10^{-3}\,\mathrm{m\,K}$

  where λ_{max} is the wavelength of maximum intensity, the *peak wavelength*.

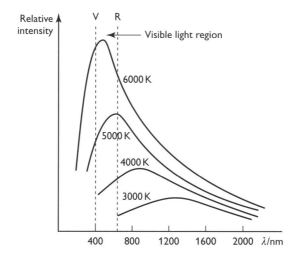

The spectral emission from a black body for a range of temperatures is illustrated in the diagram above. It can be seen that the peak wavelength shifts towards lower wavelengths as the temperature increases. This can be simply demonstrated by heating a charcoal block with a Bunsen burner flame. After a short time the block still appears black, but if the flame is removed the radiant heat of the infrared emissions can be felt by placing your hand close to the block. After further heating, the block glows red and, as the temperature rises, orange, then yellow; finally the block glows white hot.

Wien's law is derived from these curves. They were also used by Max Planck to establish the foundations of quantum theory (see Units 2 and 4). At higher temperatures there are more energetic photons, so the equation $\Delta E = hf = \dfrac{hc}{\lambda}$ is consistent with the shorter peak wavelengths at higher temperatures.

Worked example

The Sun has a radius of 6.90×10^8 m and a surface temperature of 5800 K. Gamma Crucis has a radius of 7.90×10^{10} m and a surface temperature of 3400 K.

(a) Calculate the luminosity of the Sun and of Gamma Crucis.

(b) Calculate the peak wavelength of the Sun and of Gamma Crucis.

(c) Use your answers to describe the difference in appearance of the two stars, and state their positions on the Hertzsprung–Russell diagram.

Answer

(a) $L = 4\pi r^2 \sigma T^4$

Sun: $L = 4\pi \times (6.90 \times 10^8 \text{ m})^2 \times (5.67 \times 10^{-8} \text{ W m}^{-2}\text{K}^{-4}) \times (5800 \text{ K})^4$

$\qquad = 3.84 \times 10^{26} \text{ W}$

Gamma Crucis: $L = 4\pi \times (7.90 \times 10^{10} \text{ m})^2 \times (5.67 \times 10^{-8} \text{ W m}^{-2}\text{K}^{-4}) \times (3400 \text{ K})^4$

$\qquad\qquad\qquad = 5.94 \times 10^{29} \text{ W}$

(b) $\lambda_{max}T = 2.898 \times 10^{-3} \text{ m K}$

Sun $\lambda_{max} = \dfrac{2.898 \times 10^{-3} \text{ m K}}{5800 \text{ K}}$

$\qquad = 4.99 \times 10^{-7} \text{ m} = 499 \text{ nm}$

Gamma Crucis $\lambda_{max} = \dfrac{2.898 \times 10^{-3} \text{ m K}}{3400 \text{ K}}$

$= 8.52 \times 10^{-7} \text{ m}$

$= 852 \text{ nm}$

(c) Gamma Crucis has a peak wavelength that is just beyond the red end of the visible spectrum, so most of the visible light will be in the red region. The peak wavelength for the Sun is just less than the mid-point of the spectrum. It appears whitish orange.

The Sun is a main sequence star, and Gamma Crucis, with its massive luminosity and red colour, is a red giant.

The matter thrown out by planetary nebulae and supernovae will eventually form clusters in space. If the mass is large enough, the process of star formation will begin again. The nuclear reactions in even the largest stars cannot produce elements heavier than iron. The larger elements, such as gold and uranium, are produced in stars by neutron capture. Neutrons are one of the products of fusion reactions and, because they have no charge, they readily collide with other nuclei and are captured to form new isotopes.

The expanding universe

Edwin Hubble used observations on the Andromeda nebula to show that it was in fact a congregation of stars much deeper in space than the other stars visible at that time. He concluded that the increase in wavelength of the characteristic lines of the absorption spectra of elements was a consequence of the galaxy moving away from us.

The displacement of the spectral lines towards the red end of the visible spectrum, the **redshift**, can be explained as a similar phenomenon to the Doppler effect (see Unit 2). The wavelength of radiation observed from a source moving away from us is increased (the frequency decreases), and the wavelength from a source moving towards us is decreased.

The redshift, z, is defined as the ratio of the change in wavelength to that of a stationary source (like the Sun):

$$z = \frac{\Delta\lambda}{\lambda}$$

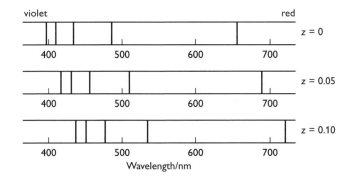

The diagram above compares some of the spectral lines of hydrogen from the Sun ($z = 0$), with those for galaxies with redshifts of 0.05 and 0.10. They show that the pattern remains the same but is shifted towards the red end of the spectrum.

For sources moving with speeds much less than the speed of light ($v \ll c$):

$$z = \frac{\Delta\lambda}{\lambda} \approx \frac{\Delta f}{f} \approx \frac{v}{c}$$

Worked example

In the laboratory, a certain spectral line in the hydrogen spectrum is observed to have a wavelength of 696 nm. The same line in the spectrum is measured by an astronomer observing the galaxy Virgo as 724 nm. Calculate the recessional speed of the galaxy.

Answer

$$\frac{\Delta\lambda}{\lambda} = \frac{v}{c} \Rightarrow v = \frac{\Delta\lambda}{\lambda} c$$

$$v = \frac{(724 - 696) \times 10^{-9}\,\text{m}}{696 \times 10^{-9}\,\text{m}} \times 3.00 \times 10^{8}\,\text{m s}^{-1}$$

$$= 1.21 \times 10^{7}\,\text{m s}^{-1}$$

$$= 12\,000\,\text{km s}^{-1}$$

Hubble's law

Hubble's observations showed that the more distant galaxies were moving away at greater speeds, and that the recessional speed, v, was approximately proportional to the distance, d, of the galaxy from Earth:

$v = H_0 d$

where H_0 is the Hubble constant, which has a value in the range 50–$100\,\mathrm{km\,s^{-1}\,Mpc^{-1}}$.

The value of the Hubble constant is variable because of the difficulty in accurately measuring the distances of galaxies that are as much as hundreds of millions of light years away from Earth. A value of $70\,\mathrm{km\,s^{-1}\,Mpc^{-1}}$ is often used.

Worked example

If the Hubble constant is $70\,\mathrm{km\,s^{-1}\,Mpc^{-1}}$ and $1\,\mathrm{Mpc}$ is $3.09 \times 10^{22}\,\mathrm{m}$, use Hubble's law to estimate the age of the universe.

Answer

Suppose a galaxy has been travelling at a speed v for a time t since the Big Bang. The distance travelled will be vt.

$$v = H_0 d$$
$$= H_0 vt$$
$$\Rightarrow t = \frac{1}{H_0}$$

Now, $H_0 = \dfrac{70 \times 10^3\,\mathrm{m\,s^{-1}}}{3.09 \times 10^{22}\,\mathrm{m}}$

$$= 2.27 \times 10^{-18}\,\mathrm{s^{-1}}$$

So $t = \dfrac{1}{2.27 \times 10^{-18}\,\mathrm{s^{-1}}}$

$$= 4.41 \times 10^{17}\,\mathrm{s}$$

$$= \frac{4.41 \times 10^{17}\,\mathrm{s}}{365.25 \times 24 \times 60 \times 60\,\mathrm{s\,y^{-1}}}$$

$$= 1.4 \times 10^{10}\ \text{years}$$

A simple model of an expanding universe uses a balloon with small swirls, representing galaxies, drawn onto the surface. When the balloon is inflated, the galaxies move apart so that, wherever you are, all the other galaxies move away from you as the balloon grows larger.

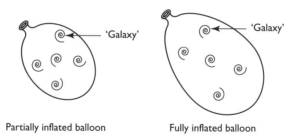

| Partially inflated balloon | Fully inflated balloon |

Whether the universe will carry on expanding is a matter for debate. As the galaxies move outwards, their gravitational potential energy increases, and consequently their kinetic energy is reduced — the rate of the expansion is therefore continually decreasing. The fate of the universe depends on its average density. If this exceeds a critical value then the gravitational energy will eventually be sufficient to reverse the expansion, and the universe will contract (a **closed universe**). If the density is less than the critical value then the universe will carry on expanding (an **open universe**).

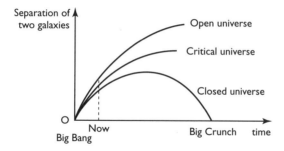

The ratio average density:critical density is given the symbol Ω. If $\Omega < 1$ the universe will be open, and if $\Omega > 1$ it will be closed.

Dark matter

Put simply, dark matter is that which cannot be detected by either its emission or absorption of radiation. Gravitational effects on the rotation of galaxies, and the gravitational lensing effect of clusters of galaxies, suggest that as much as 90% of the mass of the galaxies is dark matter.

The nature of dark matter is uncertain. It may consist of baryons, neutrinos or other undetected weakly interacting massive particles (WIMPS). Without taking dark matter into account Ω would be about 0.1, but the estimates of the dark matter in the universe suggests a density about that of the critical value, making $\Omega = 1$.

Nuclear binding energy

A nucleus consists of protons and neutrons. In the earlier section on radioactive decay, the N–Z diagram showed that the ratio of neutrons to protons increases as the proton number rises — this helps with the stability of nuclei. The protons in the nucleus are positively charged and will experience a mutual electrostatic repulsive force. They must therefore be held in place by the **strong nuclear force** when they are close together.

If a nucleus were to be assembled from its constituent nucleons, energy would be released as the nuclear forces pulled in the protons and neutrons.

- **Binding energy** is the energy released when a nucleus is formed from its constituent nucleons.

It follows that the same amount of energy would be needed to separate the nucleus into the individual protons and neutrons — this gives an alternative definition of binding energy. The loss in energy is equivalent to a loss in mass, the **mass defect**, Δm, which is represented by the expression:

$\Delta E = c^2 \Delta m$

Because mass defects are extremely small, it is often convenient to use the **unified mass unit**, u, where $1u \equiv 1.66 \times 10^{-27}\,\text{kg}$.

Worked example

Calculate the binding energy of a helium nucleus using the following data:

mass of helium nucleus = 4.000602 u
mass of proton = 1.007276 u
mass of neutron = 1.008665 u

Answer

The helium nucleus contains two protons and two neutrons.
$\Delta m = (2 \times 1.007276\,\text{u}) + (2 \times 1.008665\,\text{u}) - 4.000602\,\text{u}$
$\quad = 0.03128\,\text{u}$
$\quad = 0.03128\,\text{u} \times 1.66 \times 10^{-27}\,\text{kg}\,\text{u}^{-1}$
$\quad = 5.20 \times 10^{-29}\,\text{kg}$

Tip Don't round up the masses when using unified mass units.

$\Delta E = c^2 \Delta m$
$\quad = (3.00 \times 10^8\,\text{m}\,\text{s}^{-1})^2 \times 5.20 \times 10^{-29}\,\text{kg}$
$\quad = 4.68 \times 10^{-12}\,\text{J}$

The stability of nuclei depends on the binding energy *per nucleon*. This is usually represented in MeV per nucleon. The binding energy per nucleon varies with the nucleon number for stable elements as shown below.

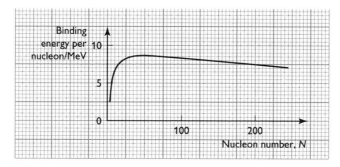

It can be seen that the binding energy per nucleon rises with nucleon number until it reaches a maximum at iron, which is the most stable nucleus. The binding energy per nucleon then falls gradually for the heavier elements.

Worked example

The nuclear masses for a range of isotopes are given below:

$^{14}_{7}N = 14.00307\,u$; $^{16}_{8}O = 15.99492\,u$;

$^{56}_{26}Fe = 55.93493\,u$; $^{206}_{82}Pb = 205.97447\,u$; $^{238}_{92}U = 238.05082\,u$.

(a) Calculate the binding energy per nucleon for each isotope, given that the mass of a proton is $1.007276\,u$, and the mass of a neutron is $1.008665\,u$. Express your answers in MeV per nucleon.

(b) Explain which of the isotopes is the most stable.

Answer

(a) Nitrogen:

$\Delta m = (7 \times 1.007276\,u) + (7 \times 1.008665\,u) - 14.00307\,u$

$\qquad = 0.108517\,u$

$\qquad = 0.108517\,u \times 1.66 \times 10^{-27}\,kg\,u^{-1}$

$\qquad = 1.80 \times 10^{-28}\,kg$

$\Delta E = (3.0 \times 10^{8}\,m\,s^{-1})^{2} \times 1.80 \times 10^{-28}\,kg$

$\qquad = 1.62 \times 10^{-11}\,J$

$\qquad = \dfrac{1.62 \times 10^{-11}\,J}{1.6 \times 10^{-13}\,J\,MeV^{-1}}$

$\qquad = 101\,MeV$

$\dfrac{\Delta E}{A} = \dfrac{101\,MeV}{14}$

$\qquad = 7.21\,MeV$ per nucleon

Oxygen:

$\Delta m = 0.132608\,u = 2.20 \times 10^{-28}\,kg$

$\Delta E = 1.98 \times 10^{-11}\,J = 124\,MeV$

$\dfrac{\Delta E}{A} = \dfrac{124\,MeV}{16} = 7.75\,MeV$ per nucleon

Iron:

$\Delta m = 0.514\,196\,u = 8.54 \times 10^{-28}\,kg$

$\Delta E = 7.68 \times 10^{-11}\,J = 480\,MeV$

$\dfrac{\Delta E}{A} = \dfrac{480\ MeV}{56} = 8.57\,MeV$ per nucleon

Lead:

$\Delta m = 1.696\,622\,u = 2.82 \times 10^{-27}\,kg$

$\Delta E = 2.53 \times 10^{-10}\,J = 1580\,MeV$

$\dfrac{\Delta E}{A} = \dfrac{1580\ MeV}{206} = 7.67\,MeV$ per nucleon

Uranium:

$\Delta m = 1.883\,662\,u = 3.13 \times 10^{-27}\,kg$

$\Delta E = 2.81 \times 10^{-10}\,J = 1760\,MeV$

$\dfrac{\Delta E}{A} = \dfrac{1760\ MeV}{238} = 7.39\,MeV$ per nucleon

(b) Iron has the largest amount of binding energy per nucleon, and is therefore the most stable of the isotopes.

Nuclear fusion

In **nuclear fusion**, energy is released when two light nuclei fuse together to form a heavier nucleus. This happens on the rising part of the binding energy per nucleon curve. Mass is lost and binding energy is released, usually in the form of high-energy gamma photons and neutrinos.

In order to initiate nuclear fusion, the Coulomb repulsive forces between the positively charged particles have to be overcome. For hydrogen fusion, high densities and temperatures of about $10^7\,K$ are needed to give the nuclei sufficient kinetic energy to overcome these forces. Because of the high temperatures needed, these are often referred to as **thermonuclear reactions**.

In the core of a main sequence star like our Sun, a proton–proton chain reaction takes place, releasing energy and producing helium nuclei. The chain has several stages, but the overall effect can be represented as the fusion of four protons to create one helium nucleus:

$$4{}^{1}_{1}\text{H} \rightarrow {}^{4}_{2}\text{He} + 2e^{+} + 2\nu_{e} + 2\gamma$$

The positrons will annihilate with electrons to create four more gamma photons (see Unit 4):

$$2e^{+} + 2e^{-} \rightarrow 4\gamma$$

Worked example

(a) Show that the energy released by the fusion of four hydrogen nuclei to create a helium nucleus is about 27 MeV, given that the mass of $^1_1H = 1.007276\,u$ and the mass of $^4_2He = 4.000602\,u$.

(b) The Sun has a mass of about $2 \times 10^{30}\,kg$ and a life of around 10^{10} years. Estimate the loss in its mass over this time due to hydrogen fusion. You may assume that the luminosity of the Sun is $4 \times 10^{26}\,W$ and that 1 year is about $3 \times 10^7\,s$. Give your answer as a percentage of the original mass.

Answer

(a) $\Delta m = 4 \times 1.007276\,u - 4.000602\,u$

$= 0.028502\,u$

$= 0.028502\,u \times 1.66 \times 10^{-27}\,kg\,u^{-1}$

$= 4.73 \times 10^{-29}\,kg$

$\Delta E = (3.00 \times 10^8\,m\,s^{-1})^2 \times 4.73 \times 10^{-29}\,kg$

$= 4.26 \times 10^{-12}\,J$

$= \dfrac{4.26 \times 10^{-12}\,J}{1.6 \times 10^{-13}\,J\,MeV^{-1}}$

$= 26.6\,MeV$

Tip Always give answers to one more significant figure in 'show that' questions.

(b) Number of fusions per second $= \dfrac{4 \times 10^{26}\,W}{4.26 \times 10^{-12}\,J}$

$= 9 \times 10^{37}\,s^{-1}$

Mass converted $= 9 \times 10^{37}\,s^{-1} \times 10^{10}\,y \times 3 \times 10^7\,s\,y^{-1} \times 4.73 \times 10^{-29}\,kg$

$= 1.3 \times 10^{27}\,kg$

Percentage loss $= \dfrac{1.3 \times 10^{27}\,kg}{2 \times 10^{30}\,kg} \times 100\%$

$= 0.07\%$

In larger stars the core temperature can rise up to 10^8–$10^9\,K$, enabling the fusion of heavier elements to take place. It can be seen from the binding energy per nucleon curve that iron has the maximum binding energy per nucleon, and so the fusion of particles to produce heavier elements would require an additional input of energy — fusions beyond this limit are not possible.

Hydrogen fusion could be the answer to the world's energy needs. However, the problems of controlling very hot, high-density hydrogen plasma for a sustained period are insurmountable at present (temperatures of $10^7\,K$ would vaporise the walls of any solid container). Projects such as the Joint European Torus have generated short bursts of fusion energy from plasma held in powerful magnetic fields, but the technology needed for a continuous energy source is well beyond us at the moment.

Nuclear fission

In **nuclear fission**, a massive nucleus, such as that of uranium, splits into two lighter nuclei. When this happens the smaller nuclei have more binding energy per nucleon than the original uranium nuclei, so that mass is lost and energy is released.

It can be quite confusing that events on the rising part of the binding energy per nucleon curve lead to a release of energy. It may be easier to imagine a nuclear potential energy curve. For iron, the nucleons are pulled inward by the strong nuclear force to give the maximum binding energy per nucleon, but they are at the point of lowest nuclear potential energy. The potential energy against nucleon number curve will be the inverse of the binding energy curve. It becomes easier to see that when two light nuclei fuse together, or when a heavy nucleus splits into smaller ones, the potential energy falls and energy is released.

Unlike fusion reactions, high temperatures and pressures are not needed, but they are usually triggered by neutron bombardment. A typical example is the fission of uranium-235 used to generate energy in many fission reactors:

$$^{235}_{92}U + {}^{1}_{0}n \rightarrow {}^{236}_{92}U \rightarrow {}^{141}_{56}Ba + {}^{92}_{36}Kr + 3{}^{1}_{0}n$$

This reaction is typical, in that the fission products are not usually the same (mass ratios of the order of $3:2$ are most probable) and more neutrons are emitted. The extra neutrons are a consequence of the lower neutron to proton ratios of the smaller nuclei (see the section on nuclear stability) — these will initiate further fissions, resulting in a chain reaction.

Worked example

(a) Use the data in the table to show that the energy released from the fission of a single nucleus of uranium-235 in the reaction given above, is about 3×10^{-11} J.

Particle	${}^{1}_{0}n$	${}^{235}_{92}U$	${}^{141}_{56}Ba$	${}^{92}_{36}Kr$
Mass/u	1.008665	235.04393	140.91434	91.92625

(b) Calculate the energy released by the fission of 1 kg of uranium-235, and estimate the annual consumption of the fuel in a 250 MW power station. State any assumptions that you make.

Answer

(a) initial mass = 235.04393 u + 1.008665 u

$\quad\quad\quad\quad\quad$ = 236.052595 u

Final mass = 140.91434 u + 91.92625 u + (3 × 1.008665 u) = 235.866585 u

Δm = 236.052595 u − 235.866585 u

$\quad\quad$ = 0.18601 u

$\quad\quad$ = 0.18601 u × 1.66 ×10^{-27} kg u^{-1}

$\quad\quad$ = 3.09 × 10^{-28} kg

ΔE = (3.00 × 10^{8} m s^{-1})^2 × 3.09 × 10^{-28} kg = 2.78 × 10^{-11} J

(b) 1 kg of uranium-235 contains $\dfrac{1\text{kg}}{235 \times 1.66 \times 10^{-27}\text{kg}} = 2.56 \times 10^{24}$ nuclei.

The energy released per kg $= 2.56 \times 10^{24} \times 2.78 \times 10^{-11}\text{J}$

$$= 7.1 \times 10^{13}\text{Jkg}^{-1}$$

Assuming that the efficiency of the power station is 25%, the power generated in the reactor will be 1000 MW, and 1 year $= 3 \times 10^7\text{s}$

Energy generated in 1 year $= 1000 \times 10^6\text{W} \times 3 \times 10^7\text{s}$

$$= 3 \times 10^{16}\text{J}$$

Mass of uranium-235 used in 1 year $= \dfrac{3 \times 10^{16}\,\text{J}}{7 \times 10^{13}\text{J kg}^{-1}} = 400\text{kg}$

Note: Naturally occurring uranium contains only 0.7% of uranium-235 (although it is enriched to 3−5% for reactor fuel) so around 60 tonnes would be needed each year for such a reactor. This is very small compared with the millions of tons of coal needed for a similar coal-fired power station.

The advantages of nuclear fission power are that relatively small masses of fuel are required and that virtually no greenhouse gases are emitted. These have to be balanced against the environmental problems of the storage of fission products (some of which have very long half-lives) and the production of fissile material suitable for nuclear weaponry.

Questions
&
Answers

The following two tests are made up of questions similar in style and content to those appearing in a typical Unit 5 examination. Each paper carries a total of 80 marks and should be completed in 1 hour 35 minutes. The first ten questions are multiple-choice objective tests with four alternative responses each. The remaining questions vary in length and style, and the number of marks that each is worth ranges from 3 to 14. You might like to work through an entire paper in the allotted time and then check your answers; alternatively, you could separately attempt the multiple-choice section and selected longer questions to fit your revision plan. Use the fact that there are 95 minutes available for the 80 marks on the test to help you estimate how long you ought to spend on a particular question — you should be looking at about 10 minutes for the multiple-choice section, and approximately 5 minutes for a 4 mark question and 15 minutes for one carrying 12 marks.

Although these sample papers resemble actual examination scripts in most respects, be aware that during the examination you will be writing your answers directly onto the paper, which is not possible for the tests in this book. It may be that you will need to copy diagrams and graphs that you would normally just write or draw onto in the real examination. If you are attempting one of these papers as a timed test, allow yourself an extra few minutes to account for this.

The data sheet should be used for these tests. Values for physical quantities such as the mass of an electron will not be shown in the questions, and many important relationships can be found on the data sheet. You should spend some time familiarising yourself with them.

The answers should not be treated as model solutions because they represent the bare minimum necessary to gain the marks. In some instances, the difference between an A-grade response and a C-grade response is suggested. This is not possible for the multiple-choice section, and many of the shorter questions do not require extended writing.

For question parts worth multiple marks, ticks ✓ are included in the answers to indicate where the examiner has awarded a mark. Half marks are not given.

Examiner's comments

Where appropriate, the answers are followed by examiner's comments, denoted by the icon 🄴. These are interspersed in the answers and indicate where credit is due and where lower grade candidates typically make common errors. They may also provide useful tips.

Test Paper 1

Questions 1–10

For questions 1–10 select one answer from **A** to **D**.

(1) A valid set of units for specific heat capacity is:

    A JK^{-1}

    B Jkg^{-1}

    C $Jkg^{-1}K^{-1}$

    D $JkgK^{-1}$ (1 mark)

(2) Technetium-99m is used extensively in medical imaging primarily because:

    A it emits both alpha and beta particles

    B it has a very long half-life

    C it is a pure gamma emitter

    D it is readily absorbed by the thyroid gland   (1 mark)

(3) The surface temperature of a star that emits radiation with a peak wavelength of 600 nm is approximately:

    A 500 K

    B 2000 K

    C 5000 K

    D 20 000 K   (1 mark)

The graph shows the variation of acceleration with displacement from the equilibrium position of a body undergoing simple harmonic motion. Use the graph to answer questions 4 and 5.

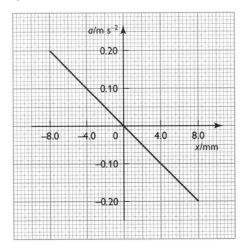

(4) The amplitude of the oscillation is:

A 8 mm

B 16 mm

C 0.20 m s^{-2}

D 0.40 m s^{-2} (1 mark)

(5) The period of the motion is:

A 0.04π

B 0.08π

C 0.2π

D 0.4π (1 mark)

(6) The gravitational field strength on the surface of the Earth is g. The gravitational field strength on the surface of a planet that has twice the mass and twice the radius of Earth is:

A $\frac{g}{4}$

B $\frac{g}{2}$

C g

D $4g$ (1 mark)

(7) The diagram shows the variation of the amplitude of a lightly damped oscillator as it is forced to vibrate over a range of frequencies.

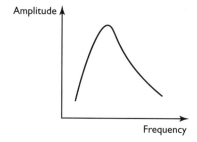

Which of the following statements best describes the change in shape of the graph when the damping is increased?

A The maximum amplitude falls and f_0 decreases

B The maximum amplitude falls and f_0 increases

C The maximum amplitude falls and f_0 stays the same

D The maximum amplitude rises and f_0 stays the same (1 mark)

(8) In astronomy, a *standard candle* is:

A a star of known luminosity

B a unit of heat

C a unit of luminosity

D a unit of radiant flux (1 mark)

Questions 9 and 10 relate to the fusion reaction:

$$4^1_1H \rightarrow {}^4_2He + 2e^+ + 2\nu_e + 2\gamma$$

(9) The energy released in this reaction if there is a *mass defect* of 5.0×10^{-29} kg is:

A 1.5×10^{-20} J

B 4.5×10^{-20} J

C 2.3×10^{-12} J

D 4.5×10^{-12} J   (1 mark)

(10) Such a reaction is most likely to take place:

A in a neutron star

B in the core of a main sequence star

C in the core of a red giant star

D in a white dwarf star   (1 mark)

Total: 10 marks

■ ■ ■

Answers to Questions 1–10

(1) C

🖉 Using the equation $\Delta Q = mc\Delta\theta \Rightarrow c = \dfrac{\Delta Q}{m\Delta\theta}$ so the units are $Jkg^{-1}K^{-1}$.

(2) C

🖉 Gamma radiation is much less ionising than alpha or beta particles. There will be less damage to tissue, and most of the radiation can be detected outside the body. Technetium-99m has a half-life of about 6 hours, a suitable time for diagnostic examination. It is not preferentially absorbed by the thyroid gland.

(3) C

🖉 Using Wien's law; $\lambda_{max}T = 2.898 \times 10^{-3}\,mK \Rightarrow T = \dfrac{2.898 \times 10^{-3}\,m\,K}{600 \times 10^{-9}\,m} \approx 5000\,K$

(4) A

🖉 Amplitude = maximum displacement about the equilibrium position = 8 mm

(5) D

🖉 Gradient $= -\omega^2 \Rightarrow \dfrac{4\pi^2}{T^2} = \dfrac{0.20\,ms^{-2}}{8 \times 10^{-3}\,m} = 25s^{-2} \Rightarrow T = \dfrac{2\pi}{5} = 0.4\pi$

(6) B

🖉 $g = \dfrac{GM}{R^2} \Rightarrow g_{planet} = \dfrac{G \times 2M}{(2R)^2} = \dfrac{g}{2}$

(7) A

🖉 When the oscillation is damped, energy is lost from the system so the amplitude falls. The damping also causes the peak frequency to be reduced.

(8) A

 A standard candle is a star of a given type, for which the luminosity and distance are known, so that the distances of similar types of star (with the same luminosity) can be measured.

(9) D

 $\Delta E = c^2 \Delta m = (3.00 \times 10^8 \, \text{m s}^{-1})^2 \times 5.0 \times 10^{-29} \, \text{kg} = 4.5 \times 10^{-12} \, \text{J}$

(10) B

 Hydrogen fusion occurs in the core of a main sequence star. Fusion of heavier elements takes place in the core of a red giant (hydrogen fusion can take place outside the core) and there is no fusion in a white dwarf (no fuel) or in a neutron star.

Question 11

(a) **Explain what is meant by the *internal energy* of a body.** (2 marks)

(b) **How does the internal energy of an ideal gas differ from that of a real gas?** (1 mark)

(c) **An ideal gas is contained in a volume of $5.0 \times 10^{-3}\,m^3$ at a temperature of 17°C. Calculate the number of molecules of gas within this container if the pressure is $1.2 \times 10^5\,Pa$** (2 marks)

Total: 5 marks

Answer to Question 11

(a) The internal energy is the sum of the kinetic energy of the particles ✓ and the potential energy ✓ in the bonds between the particles.

🖉 A C-grade candidate may be aware that the internal energy relates to the energy of the molecules without stating both potential and kinetic energy.

(b) In an ideal gas there are no forces between the particles ✓ so they have only kinetic energy.

(c) $pV = NkT \Rightarrow N = \dfrac{pV}{kT}$

$$N = \frac{1.2 \times 10^5\,Pa \times 5.0 \times 10^{-3}\,m^3}{1.38 \times 10^{-23}\,J\,K^{-1} \times (273+17)K}$$

$$= 1.5 \times 10^{23}$$

The marks are awarded for correct use of formula ✓; conversion to K plus correct answer ✓.

Question 12

A mass oscillating on a spring is an example of simple harmonic motion.

(a) State the conditions required for simple harmonic motion to take place. (2 marks)

The graph shows how the kinetic energy varies with displacement for a particular mass and spring.

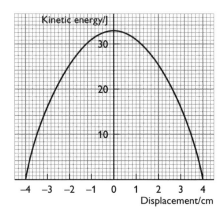

(b) Use the graph to calculate:

    (i)   the total energy of the system (1 mark)

    (ii) the potential energy of the mass when the displacement is 2.0 cm (2 marks)

    (iii) the stiffness, k, of the spring. (3 marks)

Total: 8 marks

■ ■ ■

Answer to Question 12

(a) The resultant force (or acceleration) must be proportional to the displacement ✓ and directed towards the equilibrium position ✓.

e A grade-C candidate may state that the acceleration is directed towards a fixed point, without specifying equilibrium, and would gain only a single mark.

(b) (i) $E_{\text{total}} = 32\,\text{J}$ ✓

    (ii) $E_p = E_{\text{total}} - E_k$ ✓ $= 32\,\text{J} - 24\,\text{J} = 8\,\text{J}$ ✓

    (iii) $E_p = \frac{1}{2}kx^2$ ✓

$$k = \frac{2E_p}{x^2} = \frac{2 \times 8\,\text{J}}{(2.0 \times 10^{-2}\,\text{m})^2} \quad \text{(or with any valid readings from the graph)} \checkmark$$

$$= 4.0 \times 10^4\,\text{N}\,\text{m}^{-1} \checkmark$$

Question 13

(a) **Explain the meaning of** *background radiation*. **You should include two sources of the radiation in your explanation.** (2 marks)

(b) **Suggest a simple experiment to estimate the percentage of gamma rays in the background radiation.** (2 marks)

Total 4 marks

■ ■ ■

Answer to Question 13

(a) Background radiation is that which occurs naturally and is always present in the environment ✓; sources of background radiation are cosmic rays, rocks, food and man-made effects due to weapon-testing and waste from nuclear reactors; any two ✓.

(b) Use a detector (e.g. Geiger counter) to measure the background radiation. Cover the input with a piece of aluminium a few millimetres thick and take the reading again ✓; alpha and beta particles will be absorbed by the aluminium, so only the gamma radiation will be registered ✓.

Question 14

(a) **State one similarity and one difference between electric fields and gravitational fields.** (2 marks)

(b) **Two electrons are 1 metre apart in a vacuum. Calculate: (i) the gravitational force, (ii) the electrostatic force between them.** (4 marks)

Total: 6 marks

■ ■ ■

Answer to Question 14

(a) Similarity: inverse square laws (or Coulomb and Newton laws given); g is force per unit mass, E is force per unit charge; units: $N\,kg^{-1}$ and $N\,C^{-1}$; any one ✓.

Difference: gravitational forces act on masses, electric forces act on charges; gravitational forces can only be attractive, electric can be attractive or repulsive; any one ✓.

(b) (i) $F = \dfrac{Gm_1 m_2}{r^2}$ ✓

$$= \frac{6.67 \times 10^{-11}\,N\,m^2\,kg^{-2} \times \left(9.11 \times 10^{-31}\,kg\right)^2}{\left(1\,m\right)^2}$$

$= 5.54 \times 10^{-71}\,N$ ✓

(ii) $F = \dfrac{kQ_1 Q_2}{r^2}$ ✓

$$= \frac{8.99 \times 10^{9}\,N\,m^2\,C^{-2} \times \left(1.6 \times 10^{-19}\,C\right)^2}{\left(1\,m\right)^2}$$

$= 2.30 \times 10^{-28}\,N$ ✓

Question 15

Part of the uranium-238 series includes the decay of thorium-234 to protactinium-234 by β^- emission, which subsequently decays to an isotope of uranium. The nuclear equations are given below:

$$^{234}_{90}\text{Th} \rightarrow {}^{234}\text{Pa} + \beta^-$$
$$^{234}\text{Pa} \rightarrow \text{U} + \beta^-$$

(a) Complete the equations by adding the appropriate values of proton number and nucleon number to the protactinium and uranium isotopes and the beta-minus particles. (2 marks)

It is possible to isolate a sample of the protactinium by dissolving it in an organic solvent, and determine its half-life by monitoring its activity.

(b) Explain the meaning of *half-life*. (1 mark)

A set of values for the activity of a sample of protactinium-234 (corrected for background radiation) taken at 30-second intervals is given in the table below.

t/s	0	30	60	90	120	150	180
A/s^{-1}	200	150	117	91	68	50	40

The relationship between the activity and the time is given by the equation:

$A = A_0\,e^{-\lambda t}$

(c) Plot a suitable graph to determine the value of the decay constant, λ, for protactinium-234. You will need to copy the table and add any further rows of deduced values that may be needed in order to complete your graph. (5 marks)

(d) Use your graph to determine the value of the decay constant, and hence calculate the half-life of the protactinium-234. (4 marks)

(e) The uranium-234 produced decays by alpha emission. Why is this unlikely to affect the activity readings? (1 mark)

Total: 13 marks

■ ■ ■

Answer to Question 15

(a) $^{234}_{90}\text{Th} \rightarrow {}^{234}_{91}\text{Pa} + {}^{0}_{-1}\beta$ ✓; $^{234}_{91}\text{Pa} \rightarrow {}^{234}_{92}\text{U} + {}^{0}_{-1}\beta$ ✓

(b) Half-life is the average time taken for the activity of a radioisotope to decay to one-half of its initial value ✓.

🖉 It is essential to include the *average* time, an omission that often loses marks in examinations. This is because of the random nature of nuclear decay.

(c)

t/s	0	30	60	90	120	150	180
A/s⁻¹	200	150	117	91	68	50	40
$\ln(A/s^{-1})$	5.30	5.01	4.76	4.51	4.22	3.91	3.67

Extra row with values of ln (A/s^{-1}) ✓.

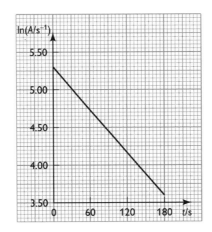

Marks are awarded for: axes correctly labelled with units ✓; scale covering at least two thirds of available squares ✓; plots correctly plotted ✓; line of best fit drawn ✓.

✍ For data handling questions it is possible to gain the graph marks even if an inappropriate plot is made. In this instance, a graph of activity against time could gain 4 marks.

(d) $\ln A = \ln A_0 - \lambda t$ ✓

Gradient of line = $\dfrac{3.67 - 5.30}{180s}$ ✓

$\lambda = 9.1 \times 10^{-3} s^{-1}$ ✓; half-life = $\dfrac{\ln 2}{9.1 \times 10^{-3} s^{-1}} = 76 s$ ✓

✍ A grade A candidate will usually draw a large triangle on the graph to determine the gradient, and also be aware that the gradient has a negative value.

(e) The alpha particles emitted by the uranium-234 will be absorbed by the container and so will not affect the activity readings ✓.

Question 16

A Hertzsprung–Russell diagram plots the ratio of the luminosity of a number of known stars to that of the Sun ($L_\odot = 3.9 \times 10^{26}\,$W) against the star's surface temperature.

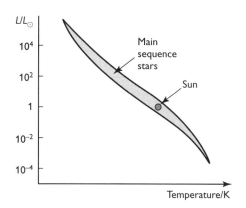

The region drawn on the diagram represents the main sequence stars.

(a) What is a *main sequence* star? (2 marks)

(b) Complete the diagram by adding:
   (i)   a suitable temperature scale
   (ii) the regions in which the red giants (label **R**) and the white dwarfs (label **W**) are located. (4 marks)

(c) The luminosity scale is logarithmic. Explain why such a scale is used. (1 mark)

(d) With reference to the HR diagram, describe the changes that happen to a star similar to the Sun when it leaves the main sequence to become a red giant. (5 marks)

(e) A white dwarf will eventually fade away and disappear from view. Explain why this happens. (2 marks)

Total: 14 marks

■ ■ ■

Answer to Question 16

(a) A main sequence star maintains a constant temperature and luminosity by the fusion of hydrogen to form helium ✓ in its core ✓.

✎ It is important to state that hydrogen fuses to form helium, *and* that this takes place in the core. A common error is the omission of the fusion product and/or that hydrogen fusion occurs 'within the star'.

(b)

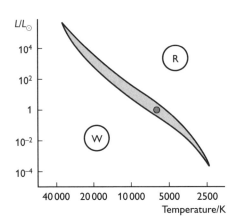

(i) Marks are awarded for: scale increasing from right to left ✓; logarithmic and minimum range 4000–30 000 K ✓.

(ii) Marks are awarded for: W left of Sun and lower than $L = 1$ ✓; R right of Sun and higher than $L = 1$ ✓.

(c) The scale is logarithmic in order to include a wide range of luminosities ✓.

(d) Hydrogen fusion ceases/hydrogen used up (in the core) ✓.

The core collapses ✓.

Hydrogen fusion initiated in the shell ✓.

The shell expands ✓.

The temperature falls and the luminosity rises ✓.

📝 The final mark can be gained only if reference is made to the relative position of the red giant to the Sun on the HR diagram — as required by the question.

(e) Fusion has ceased in the white dwarf ✓ so it will cool down ✓ and eventually fade away.

Question 17

In June 2000, a footbridge was opened across the Thames in London, to commemorate the Millennium. The bridge was closed three days later because of excessive swaying when people walked over it.

Initially it was suggested that positive feedback from the oscillations made the pedestrians walk 'in step', causing the bridge to resonate. The problem was solved by incorporating about 90 extra dampers into the structure, and the bridge was re-opened in 2002.

Recent research suggests that the sway was unlikely to have been due to pedestrians walking at the same frequency, but by the negative damping created by the lateral forces of the pedestrians' feet as they adjusted their balance.

With reference to the above passage, explain the terms *resonance*, *dampers* and *negative damping*.

Total: 6 marks

■ ■ ■

Answer to Question 17

Resonance is the increase in amplitude of the oscillations of the bridge ✓ when the driver frequency is the same as its natural frequency ✓.

Dampers are devices that absorb energy from the swaying bridge ✓ and so reduce the amplitude ✓.

Negative damping is the energy input ✓ from the pedestrians' feet ✓.

🖉 Once again, the question requires you to refer to the passage. A grade-C candidate may lose marks by giving valid definitions without reference to the bridge.

Question 18

The diagram shows a graph of the binding energy per nucleon against nucleon number for all the elements.

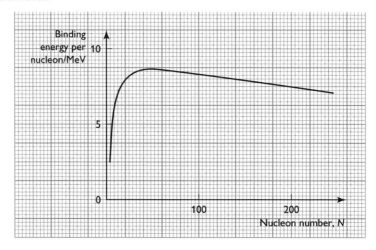

(a) What is meant by binding energy? (1 mark)

A typical fission reaction that takes place in a nuclear reactor is represented by the expression:

$$^{235}_{92}U + {}^{1}_{0}n \rightarrow {}^{236}_{92}U \rightarrow {}^{141}_{56}Ba + {}^{92}_{36}Kr + 3{}^{1}_{0}n$$

(b) Use the graph to estimate the binding energy (in MeV) of uranium-235, barium-141 and krypton-92. (3 marks)
(c) Show that the energy released by the fission of one nucleus of uranium-235 is about 3×10^{-11} J. (3 marks)

Total: 7 marks

■ ■ ■

Answer to Question 18

(a) Binding energy is the energy that would be released when separate nucleons combine to form a nucleus (or the energy needed to split a nucleus into separate nucleons) ✓.

(b) From the graph, the binding energy per nucleon is:

uranium-235 = 7.2 (±0.2) MeV

barium-141 = 8.0 (±0.2) MeV

krypton-92 = 8.4 (±0.2) MeV any two ✓ ✓

The binding energies are:

U-235 = 235 × 7.2 ≈ 1700 MeV

Ba-141 = 141 × 8.0 ≈ 1100 MeV

Kr-92 = 92 × 8.4 ≈ 800 MeV any one converted to MeV ✓

Many candidates lose marks by failing to multiply the binding energy per nucleon by the nucleon number to get the total binding energy in the nucleus.

(c) Energy released = (1100 + 800) MeV − 1700 MeV = 200 MeV ✓

$$= 200 \, \text{MeV} \times 1.6 \times 10^{-13} \, \text{J MeV}^{-1} \checkmark$$

$$= 3.2 \times 10^{-11} \, \text{J} \checkmark$$

Question 19

Television and communication satellites are maintained in geostationary orbits around the Earth directly above the Equator.

(a) What is meant by a *geostationary orbit*? (1 mark)

(b) Write an expression for the gravitational force acting on a satellite of mass m, orbiting the Earth, mass M, in a circular orbit of radius r. (1 mark)

(c) Use this, together with an expression for the centripetal force required to maintain such an orbit, to show that the radius is given by:

$$r^3 = \frac{GMT^2}{4\pi^2}$$

where T is the time taken to complete one orbit. (3 marks)

(d) Calculate the radius of a geostationary orbit, given that the mass of the Earth, M, is 6.0×10^{24} kg. (2 marks)

Total: 7 marks

Total for paper: 80 marks

■ ■ ■

Answer to Question 19

(a) A geostationary orbit is such that the satellite in it maintains a fixed position relative to the Earth ✓.

(b) $F = \dfrac{GMm}{r^2}$ ✓

(c) Centripetal force $= m\omega^2 r = m\dfrac{4\pi^2}{T^2}r$ ✓ $= \dfrac{GMm}{r^2}$ ✓

$\Rightarrow r^3 = \dfrac{GMT^2}{4\pi^2}$ ✓

(d) $r^3 = \dfrac{6.67 \times 10^{-11} \text{ N m}^2 \text{kg}^{-2} \times 6.0 \times 10^{24} \text{kg} \times (24 \times 60 \times 60 \text{s})^2}{4\pi^2}$ ✓

$\Rightarrow r = 4.2 \times 10^7 \text{m}$ ✓

Test Paper 2

Questions 1–10

For questions 1–10 select one answer from A to D.

(1) The temperature of an ideal gas is *not* affected by changes in:

    A the average kinetic energy of the molecules

    B the average potential energy of the molecules

    C the average speed of the molecules

    D the internal energy of the gas   (1 mark)

Carbon nuclei can be fused together to create a neon nucleus. Questions 2 and 3 relate to the nuclear equation of such a reaction:

$$^{12}_{6}C + \,^{12}_{6}C \rightarrow \,^{20}_{10}Ne + X$$

(2) The product **X** is a:

    A helium nucleus

    B neutron

    C proton

    D X-ray   (1 mark)

(3) This reaction is most likely to occur in:

    A a neutron star

    B a nuclear reactor

    C the core of a massive star

    D the Sun   (1 mark)

(4) The luminosity of a star is:

    A the power per unit area emitted from its surface

    B the power per unit area received on Earth

    C the total power emitted by the star

    D the total power reaching the Earth from the star   (1 mark)

(5) Americium-241 is a suitable isotope for use in smoke detectors because:

    A it decays by alpha emission

    B it decays by gamma emission

    C it has a very short half-life

    D it is more active at higher temperatures   (1 mark)

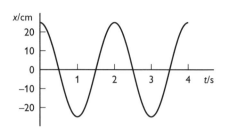

The graph shows the variation of displacement with time of a mass oscillating on a spring. Use the graph to answer questions 6 and 7.

(6) The frequency of the oscillation is:
   A 0.25 Hz
   B 0.50 Hz
   C 1.0 Hz
   D 2.0 Hz   (1 mark)

(7) The maximum velocity of the mass in m s^{-1} is:
   A 0.25π
   B 0.50π
   C π
   D 2.0π (1 mark)

(8) Which of the following does *not* explain the existence of dark matter in the universe?
   A Gravitational lensing
   B Interstellar dust
   C Random motion of galaxies
   D Rotation of galaxies   (1 mark)

The diagram shows the positions of three stars on a Hertzsprung–Russell diagram.

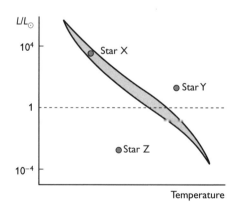

In questions **9** and **10** choose the response that best describes the selected star.

(9) **Star X:**
   A has a longer lifetime than the Sun
   B is a main sequence star
   C is a red giant
   D is less massive than the Sun   (1 mark)

(10) **Star Z:**
   A consists mainly of hydrogen
   B fuses heavy elements in its core
   C is cooler than star **Y**
   D is denser than our Sun   (1 mark)

Total 10 marks

Answers to Questions 1–10

(1) B

  There are no forces between the molecules of an ideal gas, so they have no potential energy.

(2) A

  To balance the equation, the proton number must be 2 and the nucleon number 4, which describes a helium nucleus.

(3) C

  Smaller stars such as the Sun fuse only small elements, even at the red giant stage. The higher temperatures in the cores of massive stars enable larger elements to fuse. There is no fusion in a white dwarf or a nuclear reactor.

(4) C

  Luminosity is the total power emitted by a star.

(5) A

  An alpha source is needed to ionise the smoke particles. It is relatively safe because the radiation will not penetrate the walls of the container.

(6) B

  The period is 2.0 s and so the frequency $\left(= \dfrac{1}{T}\right)$ will be 0.50 Hz.

(7) A

  $v_{max} = \omega A = 2\pi \times 0.50\,\text{Hz} \times 0.25\,\text{m} = 0.25\pi\,\text{m s}^{-1}$

(8) B

 Interstellar dust may be observed as dark streaks and so must be absorbing radiation. Dark matter does not emit or absorb radiation.

(9) B

 Star X is on the main sequence section of the diagram. It is much bigger than the Sun and so has a shorter life. It is much hotter than a red giant.

(10) D

 Star Z is a white dwarf. It has exhausted its fuel and is small, hot and very dense. The temperature scale increases to the left on an HR diagram so Z is hotter than Y.

Question 11

The particles carrying a sound wave oscillate with simple harmonic motion.

(a) Describe the energy changes of a particle in the wave during one complete oscillation. (2 marks)

(b) Calculate the total energy of an oxygen molecule oscillating with amplitude 0.5 nm in a sound wave of frequency 212 Hz. The mass of an oxygen molecule is 5.3×10^{-26} kg. (3 marks)

(c) The cavity wall around a room is filled with rubber foam. Explain, in terms of the energy of the wave particles, how this will help to soundproof the room. (2 marks)

Total: 7 marks

Answer to Question 11

(a) The energy is kinetic at the equilibrium position ✓; this changes to potential energy as the displacement increases, to a maximum value at each end of the oscillation ✓.

(b) Total energy $= \frac{1}{2}m\omega^2 A^2$ ✓

$$= \frac{1}{2} \times 5.3 \times 10^{-26}\,\text{kg} \times (2\pi \times 212\,\text{Hz})^2 \times (0.5 \times 10^{-9}\,\text{m})^2 \checkmark$$

$$= 1.2 \times 10^{-38}\,\text{J} \checkmark$$

(c) The vibrating air molecules frequently collide inelastically with the rubber ✓, so energy is transferred from the vibrating molecules ✓.

A grade C candidate may state that the molecules lose energy, but may not state that the collisions are inelastic due to the nature of the rubber.

Question 12

A combination domestic boiler is designed to provide both central heating and hot water directly from the mains supply. A typical model claims to supply a continuous flow of hot water at a rate of 12 litres per minute with a temperature rise of 35°C.

(a) Show that the energy delivered to the water every second is about 30 kJ.
Specific heat capacity of water = 4200 J kg^{-1} K^{-1}; mass of 1 litre of water = 1.00 kg   (2 marks)

(b) The manufacturer claims that the boiler is 90% efficient. Calculate the maximum power input required to provide hot water at this rate.   (1 mark)

Total: 3 marks

Answer to Question 12

(a) $\dfrac{Q}{t} = \dfrac{mc\Delta\theta}{t}$ ✓

$= \dfrac{12 \text{ kg} \times 4200 \text{ J kg}^{-1} \text{ K}^{-1} \times 35 \text{ K}}{60 \text{ s}}$

$= 29.4 \text{ kJ s}^{-1}$ ✓

(b) $90\% = \dfrac{29.4 \text{ kW}}{P_{\text{in}}} \times 100\%$

$P_{\text{in}} = 33 \text{ kW}$ ✓

Question 13

(a) Estimate the average kinetic energy of nitrogen molecules in the air when the temperature is 17°C.
(2 marks)

(b) The mass of a nitrogen molecule is 4.8 × 10⁻²⁷ kg. Show that the root mean square speed of the molecules is about 500 m s⁻¹.
(2 marks)

(c) The graph shows the distribution of molecular speeds in air at 17°C. Using the same axes, sketch a curve showing the distribution of molecular speeds when the temperature is increased.
(2 marks)

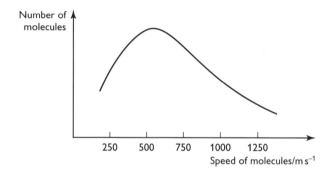

Total: 6 marks

■ ■ ■

Answer to Question 13

(a) Average kinetic energy $= \dfrac{3}{2}kT$

$= \dfrac{3}{2} \times 1.38 \times 10^{-23}\,\mathrm{J\,K^{-1}} \times 290\,\mathrm{K}$ ✓

$= 6.0 \times 10^{-21}\,\mathrm{J}$ ✓

✎ It is common for examiners to give temperatures in degrees Celsius. It is essential in all gas law equations to convert these to the Kelvin scale by adding 273 to the Celsius value.

(b) $E_k = \dfrac{1}{2}m\langle c^2 \rangle \Rightarrow \langle c^2 \rangle = \sqrt{\dfrac{2E_k}{m}}$ ✓

$\langle c^2 \rangle = \sqrt{\dfrac{2 \times 6.0 \times 10^{-21}\,\mathrm{J}}{4.8 \times 10^{-26}\,\mathrm{kg}}}$

$= 500\,\mathrm{m\,s^{-1}}$ ✓

(c)

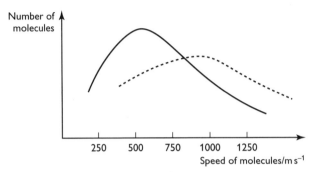

The curve peaks at a higher molecular speed ✓; with a lower maximum value ✓.

Question 14

(a) **Write an expression for the gravitational force between masses M and m separated by a distance r.** (1 mark)

(b) **Hence derive an expression for the gravitational field strength, g, at a distance r from mass M.** (2 marks)

(c) **Use the following data to show that the average density of the Moon is about 60% of that of the Earth:**

gravitational field strength on the Moon's surface = $1.6\,\mathrm{N\,kg^{-1}}$
radius of the Moon = $1.74 \times 10^6\,\mathrm{m}$
radius of the Earth = $6.37 \times 10^6\,\mathrm{m}$ (3 marks)

Total: 6 marks

■ ■ ■

Answer to Question 14

(a) $F = \dfrac{GMm}{r^2}$ ✓

(b) $g = \dfrac{F}{m}$ ✓

$= \dfrac{GMm}{r^2 m}$

$= \dfrac{GM}{r^2}$ ✓

(c) $g = \dfrac{G\frac{4}{3}\pi r^3 \rho}{r^2} = G\frac{4}{3}\pi r \rho$ ✓

🄴 You should be aware that mass = volume × density, and the volume of a sphere is included in the data at the end of the examination paper

$\Rightarrow \dfrac{\rho_M}{\rho_E} = \dfrac{1.6\,\mathrm{m\,s^{-1}} \times 6.37 \times 10^6\,\mathrm{m}}{9.8\,\mathrm{m\,s^{-1}} \times 1.74 \times 10^6\,\mathrm{m}}$ ✓

$= 0.598 = 60\%$ ✓

🄴 'Show that' questions require the answer to one more significant figure than given in the question.

Question 15

A diving bell has a volume of 8.0 m³. Initially it is suspended above the sea, in air at a temperature of 20°C and pressure 1.0 × 10⁵ Pa.

(a) Calculate the number of air molecules in the bell. (2 marks)

The bell is lowered into the sea to a depth where the air is compressed so that its pressure rises to 5.0 × 10⁵ Pa. The temperature of the water at this depth is 10°C.

Diving bell in air   Diving bell under water

(b) Calculate the volume of air in the submerged bell, stating any assumptions that you have made. (3 marks)

Total: 5 marks

Answer to Question 15

(a) $pV = NkT \Rightarrow N = \dfrac{pV}{kT}$

$$N = \frac{1.0 \times 10^5\,\text{Pa} \times 8.0\,\text{m}^3}{1.38 \times 10^{-23}\,\text{J K}^{-1} \times 293\,\text{K}} \checkmark$$

$$= 2.0 \times 10^{26} \checkmark$$

(b) Assuming that the air behaves like an ideal gas (or N or k are constant) ✓

$$\frac{p_1 V_1}{T_1} = \frac{p_2 V_2}{T_2} \quad \Rightarrow \quad V_2 = \frac{p_1 V_1 T_2}{T_1 p_2} \checkmark$$

$$V_2 = \frac{1.0 \times 10^5\,\text{Pa} \times 8.0\,\text{m}^3 \times 283\,\text{K}}{5.0 \times 10^5\,\text{Pa} \times 293\,\text{K}}$$

$$= 1.5\,\text{m}^3 \checkmark$$

e Again, be aware of the conversion from °C to K.

Question 16

In a star, the process of hydrogen burning involves protons fusing to form helium.

(a) Show that the energy released when four protons fuse to form one helium nucleus is about 4×10^{-12} J.

Masses: $^1_1p = 1.6726 \times 10^{-27}$ kg; $^4_2$He $= 6.6447 \times 10^{-27}$ kg   (3 marks)

(b) The number of protons undergoing fusion every second in the Sun is 3.8×10^{38}. Calculate the total energy emitted by the Sun every second.   (2 marks)

(c) The radiant flux reaching the Earth is $1.37 \, \text{kW m}^{-2}$. Calculate the distance of the Sun from the Earth.   (3 marks)

(d) The surface temperature of the Sun is 5800°C. Calculate the wavelength of the peak intensity radiation that it emits.   (2 marks)

(e) Sketch a graph, using the axes drawn below, to show how the intensity of the radiation received from the Sun varies with wavelength.   (2 marks)

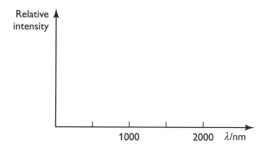

Total: 12 marks

■ ■ ■

Answer to Question 16

(a) Mass defect, $\Delta m = (4 \times 1.6726 \times 10^{-27}$ kg$) - 6.6447 \times 10^{-27}$ kg

$= 4.57 \times 10^{-29}$ kg ✓

$\Delta E = c^2 \Delta m = (3.00 \times 10^8 \, \text{m s}^{-1})^2 \times 4.57 \times 10^{-29}$ kg ✓

$= 4.1 \times 10^{-12}$ J ✓

(b) The fusion of four protons releases 4.1×10^{-12} J of energy

Energy emitted per second from the Sun $= \dfrac{3.8 \times 10^{38} \, \text{s}^{-1}}{4} \times 4.1 \times 10^{-12}$ J ✓

$= 3.9 \times 10^{26}$ W ✓

(c) $F = \dfrac{L}{4\pi d^2} \Rightarrow d = \sqrt{\dfrac{L}{4\pi F}}$ ✓

$d = \sqrt{\dfrac{3.9 \times 10^{26}\,\text{W}}{4\pi \times 1.37 \times 10^3\,\text{W m}^{-2}}}$ ✓

$= 1.5 \times 10^{11}\,\text{m}$ ✓

(d) $\lambda_{\text{max}} T = 2.898 \times 10^{-3}\,\text{m K}$

$\lambda_{\text{max}} = \dfrac{2.898 \times 10^{-3}\,\text{m K}}{6073\,\text{K}}$ ✓

$= 4.8 \times 10^{-7}\,\text{m}$

$= 480\,\text{nm}$ ✓

(e)

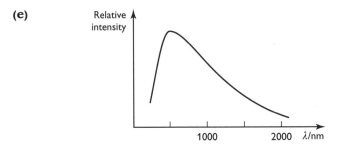

Shape ✓; peaking at approximately 500 nm ✓.

🅔 This question examines the recognition and use of basic definitions or formulae. All are included in the data sheet — always use the sheet when revising or practising examination questions.

Question 17

The table below shows the results of an experiment to investigate the relationship between the temperature of a filament lamp and the power supplied to the lamp.

V/V	I/A	P/W	log(P/W)	T/K	log(T/K)
4.0	0.61			800	
6.0	1.05			1000	
10.0	1.41			1260	
12.0	1.50			1580	

(a) Complete the table by calculating the power to the lamp, and determining the values of log(P/W) and log(T/K). (2 marks)

(b) P is related to T by a power law: $P = kT^n$.
Plot a graph of log(P/W) against log(T/K). Use your graph to determine the value of n. (6 marks)

(c) The Stefan–Boltzmann law states that the power emitted from a black body is given by the expression:

$$P = \sigma AT^4$$

Discuss the extent to which the filament lamp obeys the law. Suggest a reason why n may not be equal to 4. (2 marks)

Total: 10 marks

■ ■ ■

Answer to Question 17

(a)

V/V	I/A	P/W	log(P/W)	T/K	log(T/K)
4.0	0.61	2.4	0.38	800	2.90
6.0	0.93	5.6	0.75	1000	3.00
10.0	1.41	14.1	1.15	1260	3.10
12.0	3.00	36.0	1.56	1580	3.20

Marks are awarded for: calculations of P (= $V \times I$) ✓; logs taken for P or T ✓.

(b)

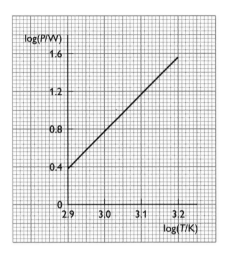

Marks are awarded for: axes covering correct range ✓; labelling with units as shown ✓; points plotted accurately ✓; line of best fit drawn ✓.

Gradient $= \dfrac{1.56 - 0.38}{3.20 - 2.90}$ or large triangle drawn ✓

$= 3.9 \, (\pm 0.1)$ ✓

📕 The mathematical requirements include the use of log–log graphs for determining the power of a quantity.

You should be aware that taking logs of the expression $P = kT^n$ leads to:

$$\log P = \log k + n \log T$$

So a graph of $\log P$ against $\log T$ will have a gradient of n (and an intercept of $\log k$).

(c) The filament probably obeys the law because 3.9 is about equal to 4 (or valid comment) ✓.

n is less than 4 due to experimental uncertainties, or because the filament is not a perfect black body ✓.

📕 In the examination, the question paper will have printed grids. Allow extra time when doing the test question in this unit guide to compensate for having to provide your own grid.

Question 18

Carbon-14 is a radioactive isotope formed in the atmosphere due to the bombardment of nitrogen nuclei neutrons by cosmic radiation. The isotope decays by beta-minus emission back to nitrogen with a half-life of 5730 years.

$$^{14}_{7}N + ^{1}_{0}n \rightarrow ^{14}_{6}C + ^{1}_{1}p$$

$$^{14}_{6}C \rightarrow ^{14}_{7}N + ^{0}_{-1}e$$

(a) What is meant by *half-life*? (1 mark)

(b) Show that the decay constant, λ, for carbon-14 is about $1.2 \times 10^{-4}y^{-1}$. (1 mark)

The ratio of carbon-14 to the stable isotope carbon-12 in the atmosphere at present is about 1.1×10^{-12}. All living organisms have this concentration in their systems, but when a plant or creature dies the carbon-14 decays and is not replenished.

(c) An oak beam from an archaeological dig has a carbon-14 to carbon-12 ratio of 4.3×10^{-13}. Estimate the age of the beam. (3 marks)

(d) Radio-carbon dating depends on the assumption that the ratio of the carbon isotopes has remained constant over the years. State two reasons why this may not be the case. (2 marks)

Total: 7 marks

■ ■ ■

Answer to Question 18

(a) Half-life is the average time taken for the activity of an isotope to fall to one half of its original value ✓.

🖉 It is essential to include *average* time in this definition. Radioactive decay is spontaneous and random so two separate measurements could differ.

(b) $\lambda = \dfrac{\ln 2}{5730 \text{ y}} = 1.21 \times 10^{-4}y^{-1}$ ✓

(c) $N = N_0 e^{-\lambda t}$

$\Rightarrow \lambda t = -\ln \dfrac{N}{N_0} = \ln \dfrac{N_0}{N}$ ✓

$1.21 \times 10^{-4}y^{-1} \times t = \ln \dfrac{1.1 \times 10^{-12}}{4.3 \times 10^{-13}}$ ✓

$\Rightarrow t = 7800$ years ✓

(d) Assumes that the ratio of carbon-14 to carbon-12 in the atmosphere has remained constant ✓ and that the half-life has stayed the same ✓.

Question 19

(a) **What is meant by 'astronomical parallax'?** (2 marks)

(b) **With the aid of a labelled diagram, describe how the distance of a nearby star can be determined using astronomical parallax.** (3 marks)

(c) **Why is this method unsuitable for measuring the distance of very distant stars?** (1 mark)

(d) **What is meant by 'redshift'? How does this support the theory of an expanding universe?** (2 marks)

The quasar 3C273 has a prominent emission line Hδ at a wavelength of 475 nm. The wavelength of same line measured from the hydrogen spectrum of a stationary source on Earth is 410 nm.

(e) **Calculate the recession velocity of 3C273.** (2 marks)

(f) **Taking the value of the Hubble constant (H_0) as $70\,\text{km}\,\text{s}^{-1}\,\text{Mpc}^{-1}$, estimate the distance of 3C273 from Earth (give your answer in Mpc).** (2 marks)

(g) **What will determine whether the universe continues to expand forever or eventually slows down and re-collapses?** (2 marks)

Total: 14 marks

Total for paper: 80 marks

■ ■ ■

Answer to Question 19

(a) 'Astronomical parallax' is the apparent movement of a nearby star against the background of distant stars ✓; when the star is observed from different positions ✓.

(b)

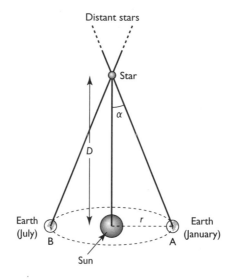

Labelled diagram ✓ ✓.

$$\alpha = \frac{r}{D} ✓$$

It is essential that a fully labelled diagram is drawn. A grade A candidate would produce a diagram as above, showing distant stars, the position of the Earth at six month intervals, and labels α, r and D. Full marks could be obtained for such a diagram plus the equation for the parallax angle. A candidate who showed no labels on the diagram would gain a maximum of 1 mark for the angle.

(c) The parallax angle is very small, even for relatively close stars. The angular displacement of distant stars against the background of more distant stars is so small that accurate measurements are not possible ✓.

(d) Redshift is the change in wavelength (or frequency) of a spectral line towards the red end of the visible spectrum, as the source moves away from the observer ✓.

The spectra from distant galaxies show redshift, so they must be moving away, and the universe is expanding ✓.

(e) $z = \dfrac{\Delta f}{f} \simeq \dfrac{\Delta \lambda}{\lambda} \simeq \dfrac{v}{c}$

$$\Rightarrow v = \frac{\Delta \lambda}{\lambda} \times c = \frac{65 \text{ nm}}{475 \text{ nm}} \times 3.00 \times 10^8 \text{ m s}^{-1} ✓$$

$$= 4.1 \times 10^7 \text{ m s}^{-1} ✓$$

(f) $v = H_0 d \Rightarrow d = \dfrac{v}{H_0} ✓$

$$d = \frac{4.1 \times 10^7 \text{ m s}^{-1}}{70 \times 10^3 \text{ m s}^{-1} \text{ Mpc}^{-1}}$$

$$= 590 \text{ Mpc} ✓$$

(g) The universe is gaining gravitational potential energy as it expands, and so will lose kinetic energy, causing the expansion to slow down ✓.

Whether it will continue to expand or collapse depends on its density ✓.

This is a 2 mark question at the end of an examination. You may be able enough to write a more detailed answer on an open or closed universe, including the effect of Ω and critical density. However, such an answer would only be expected for a discussion question carrying 5 or 6 marks. Always consider the mark allocation when answering examination questions — an overlong response could cost time and gain no extra credit.